T0176615

Statistical Analysis of Climate Extremes

The risks posed by climate change and its effect on climate extremes are an increasingly pressing societal problem. This book provides an accessible overview of the statistical analysis methods that can be used to investigate climate extremes and analyze potential risk. The statistical analysis methods are illustrated with case studies on extremes in the three major climate variables: temperature, precipitation, and wind speed. The book also provides datasets and access to appropriate analysis software, allowing the reader to replicate the case-study calculations. Providing the necessary tools to analyze climate risk, this book is invaluable for students and researchers working in the climate sciences, as well as risk analysts interested in climate extremes.

MANFRED MUDELSEE is founder of the research company Climate Risk Analysis, and a visiting scientist at the Alfred Wegener Institute, Helmholtz Centre for Polar and Marine Research. His research focuses on how climate change is related to extreme climate and weather. He is a member of the European Geosciences Union and International Association for Mathematical Geosciences.

Statistical Analysis of Climate Extremes

MANFRED MUDELSEE
Climate Risk Analysis
and
*Alfred Wegener Institute, Helmholtz Centre for Polar
and Marine Research*

CAMBRIDGE
UNIVERSITY PRESS

CAMBRIDGE
UNIVERSITY PRESS

University Printing House, Cambridge CB2 8BS, United Kingdom

One Liberty Plaza, 20th Floor, New York, NY 10006, USA

477 Williamstown Road, Port Melbourne, VIC 3207, Australia

314–321, 3rd Floor, Plot 3, Splendor Forum, Jasola District Centre,
New Delhi – 110025, India

79 Anson Road, #06–04/06, Singapore 079906

Cambridge University Press is part of the University of Cambridge.

It furthers the University's mission by disseminating knowledge in the pursuit of
education, learning, and research at the highest international levels of excellence.

www.cambridge.org
Information on this title: www.cambridge.org/9781108791465
DOI: 10.1017/9781139519441

First published 2020

Printed in the United Kingdom by TJ International Ltd, Padstow Cornwall

A catalogue record for this publication is available from the British Library.

Library of Congress Cataloging-in-Publication Data
Names: Mudelsee, Manfred, 1962- author.
Title: Statistical analysis of climate extremes / Manfred Mudelsee.
Description: New York : Cambridge University Press, 2020. | Includes
bibliographical references and index.
Identifiers: LCCN 2019040870 (print) | LCCN 2019040871 (ebook) |
ISBN 9781108791465 (paperback) | ISBN 9781139519441 (epub)
Subjects: LCSH: Climatic extremes. | Climatic extremes–Case studies. |
Climatic changes. | Climatic changes–Case studies. | Natural disasters. |
Natural disasters–Case studies.
Classification: LCC QC981.8.C53 M83 2020 (print) | LCC QC981.8.C53
(ebook) | DDC 551.6072/7–dc23
LC record available at https://lccn.loc.gov/2019040870
LC ebook record available at https://lccn.loc.gov/2019040871

ISBN 978-1-108-79146-5 Paperback

Additional resources for this publication at www.cambridge.org/mudelsee

To my family: Friederike, Tom-Luis, and Mathilda

Contents

Preface

Climate extremes cost human lives. They do harm to the economy. Examples are the Elbe flood in 2002, the European heatwave in 2003 or hurricane Katrina in 2005. The big question is how global climate change influences climate extremes. This plagues decision-makers as well as climate researchers.

This book shows how the big question can be approached using a scientific method: the statistical analysis of climate data. But the analysis is a challenge, since climate is a complex system and data are uncertain. As scientists, we have to deliver our answers with uncertainty measures (error bars) and also have to carry out computer simulations (sensitivity studies) for assessing the robustness of our results. This book shows how to do this. It is written for students, university teachers, risk analysts, and researchers.

Chapter 1 introduces you to the climate system and extremes therein. Chapter 2 informs about the various data types and methods to detect extremes. Chapter 3 details the statistical analysis tools. On the basis of this arsenal, we examine three types of climate extremes: floods and droughts (Chapter 4), heatwaves (Chapter 5), and hurricanes and other storms (Chapter 6). These examinations proceed as exemplary case studies to help you replicate the findings and carry out your own analyses. There is more information on measurements (Appendix A), climate archives (Appendix B), climate models (Appendix C), statistical inference (Appendix D), and numerical techniques (Appendix E). You can also find a description of how to access data and software (Appendix F) and a list of abbreviations and mathematical symbols (Appendix G).

To the student: Start with Chapters 1–3. Take your time to do the exercises at the end of Chapters 2 and 3. Then select from the applications (Chapters 4–6) as you please. Consult the Appendix if needed. The "Personal Reflections" about humans (including myself) and research, which are distributed throughout the book, are written for you. Consider also the "Reading Material" at the end of each chapter.

To the teacher: The main part of the book (Chapters 1–6) is designed such that it can be taught in about twenty hours. The exercises (at the end of Chapters 2 and 3) can be done as homework. The solutions to the exercises are on the website for this book. The extra time, beyond twenty hours, should be filled with computer tutorials on the replication of the book's case studies and the analysis of own data.

To the risk analyst (and other people under time pressure): Chapters 4 to 6 show in case studies how you may wish to analyze own data. The statistical analysis tools are accessible via the website for this book. Each of those chapters has at the end a section entitled "Summary for the Risk Analyst" that briefs you on the chapter's content and assesses options for near-future climate risk prediction. Consult Chapters 2 and 3 if in need of more information on data and methods, respectively.

To the researcher: Consider the case studies in Chapters 4–6 as an encouragement of your own work. Each of those chapters has at the end a section entitled "Outlook" that assesses the state of knowledge and indicates how the case studies can be extended. The given software (Appendix F) should facilitate this.

The book's website address is www.manfredmudelsee.com/textbook.

I can be electronically approached via the book's website. I would be very pleased to receive your feedback on the book: level, length, coverage, style, things to put in and things to omit. Bear in mind that humans are not perfect and that in climate science we jointly proceed to coming closer to examining the climate reality.

My work on this book benefitted from research work done previously by others including: Friedrich-Wilhelm Gerstengarbe on data from the observatory Potsdam, Michael Börngen and Gerd Tetzlaff on documentary flood data, Thomas Stocker and the rest of the IPCC on the Fifth Assessment Report, Stuart Coles on the statistical analysis of extremes and, finally, Bradley Efron and Peter Hall on the Poisson process and bootstrap resampling. I am thankful also for the influence over the past decades of my colleagues in climate research including: Gerald Haug, Gerrit Lohmann, Michael Schulz, and Martin Trauth.

Lisa Bennett was a wonderful proofreader of the first version. I thank her and also the team responsible at Cambridge University Press for their professional support and convenient communication: Sarah Lambert, Tricia Lawrence, Matt Lloyd, Zoë Pruce, and Vinithan Sedumadhavan.

Finally, I thank my family for their patience and loving support.

1
Introduction

The big question we researchers today are confronted with is whether the ongoing climate changes are associated with changes of climate extremes. Are floods occurring more often, are they getting bigger? The first part denotes the frequency aspect, the second the magnitude aspect of the question. Are we facing a higher probability of summer heatwaves in the future? Are we going to see more devastating storm events such as hurricane Katrina in August 2005? This big question plagues also politicians and stakeholders, who need to decide which measures to take to guard against the risk of those costly extreme events.

You will learn in this book how this big question can be approached using the scientific method. We researchers follow, most of us subconsciously, the philosopher Plato (427 BC–348/347 BC) and assume that there exists an objective but unknown climate reality in space and time. Key to coming closer to examining this reality is through the use of data and their mathematical analysis.

The aim of documenting data and detailing mathematical steps is to permit the reproduction of results as well as an understanding of what has been done. This allows other scientists to criticize our work by obtaining more and better data or to present improved analytical tools. The purpose is not to attack each other but to jointly come closer to the climate reality. It is inevitable that on your way to becoming a climate researcher you will meet mathematical formulas. These are things of beauty! They allow the performance of data-analytical science in the most concise way. They help us to communicate to others as effectively as possible what has been done.

Box 1.1   **Personal Reflection: Philosophy of science and irreproducible papers**

At the beginning of my PhD studies at the University of Kiel in Germany in the year 1992, I decided to put what had been my private speculations about what can be known on a more solid basis and enrolled in a subsidiary course in the philosophy of science. The influence of the teachers Hermann Schmitz and Hans Joachim Waschkies at Kiel strengthened my existing conviction of the usefulness of the working hypothesis of realism. Plato is one leading thinker on this; others are Immanuel Kant or Karl Popper. Later I found that realism corresponds to the axiomatic approach to probability, which uses data and statistical tools to infer reality with error bars.

How far should a student go beyond the boundaries of a PhD project? Is it a waste of time? My advice is to follow your curiosity, intuition, and aesthetic education. Schopenhauer's recipe is to consume original sources and not "masticated food." The philosophy books whose spines face me while I sit here have to wait until I have finished writing this textbook.

The beginning of my PhD studies brought also frustration. I wished to adapt, advance, and apply methods for analysing proxy climate time series from the seafloor archive. I browsed through papers on geology, climatology, and physics. Many of them presented fanciful tools and colorful plots – but I could not reproduce the results. The major reason for this irreproducibility was the lack of a clear description of the methods used and how the data were treated. I suffered also because I could not find a textbook that explained the ideas behind the methods in an accessible manner, or described the methods at a level that permitted the reproduction of the results. On the other hand, the philosophers I studied in my subsidiary course wrote clearly, which made their works accessible and thus criticizable.

The years subsequent to my PhD were spent writing a book on climate time series analysis (Mudelsee 2014) that aims to be accessible and clear in its descriptions. Websites with the data and the source codes of the methods can therefore make the lives of hungry students easier. The aims of accessibility and reproducibility also guide this book on analysing climate extremes.

Climate firstly can be understood as the state of the atmosphere. The Sun's incoming shortwave radiation is partly reflected by the atmosphere, partly reflected by the Earth's surface, and partly absorbed and re-emitted by the Earth as longwave radiation (Figure 1.1).

This radiative balance corresponds to a certain surface-air temperature. This variable, temperature, is the most important climate variable since it characterizes the Earth's state. Temperature interacts with many other climate variables. Relevant for the context of this book are the variables precipitation and wind speed. The radiative balance can be perturbed by changes in the atmospheric concentration of radiatively important gases, such as carbon dioxide, aerosols, or by changes in the surface properties of the Earth. These factors are the drivers of climate change (Figure 1.1).

Climate can be further understood as the state of other compartments than the atmosphere, such as the marine realms, the cryosphere (ice or snow), or the biosphere. The state in all those compartments, including the atmosphere, is not constant over space, and this spatial variability is the first reason why the climate system is complex. Furthermore, size of spatial variability depends on the climate variable. For example, it is larger for precipitation than for temperature.

The second reason for the complexity of the climate is temporal variability. Everybody experiences that day is followed by night and then another day, or that after spring there comes summer, fall, winter, and then the next spring. If these two periodic variations, the daily and the annual cycle, were the only variations, then predicting weather and climate over long time ranges would be easy. These cycles with periods of one day or one year form a part of the weather. Longer-term variations, above 30 years, form a part of the climate.

However, the climate system comprises many compartments and variables (Figure 1.1) that interact and generate a wide range of timescales of variations. We could in principle consider variations at timescales up to the age of the Earth ($\sim$4.6 Ga), the paleoclimate, but we will restrict ourselves to the climatically comparably stable Holocene (past $\sim$11.5 ka). The data from this interval (Figures 1.2 and 1.3) display a wide range of timescales of climate variations.

Climate is a complex system that varies on a wide range of scales in time and space. This means, we do not know everything about the climate, we are uncertain. Therefore, a purely mathematical apparatus is insufficient to quantitatively describe the climate. We need the statistical language. We assume that a probability (a real number between 0 and 1) can be assigned to an uncertain event, such as "it rains tomorrow" or "during the Holocene the rate of occurrence of Indian Ocean monsoon droughts more than doubled." Statistics then infers events and probabilities from the data.

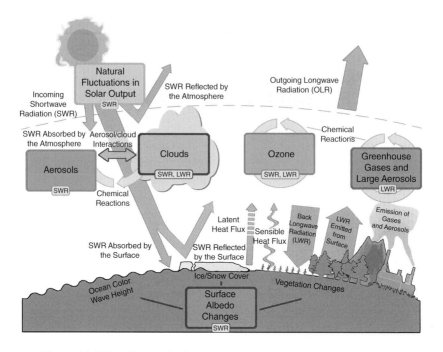

Figure 1.1 Main drivers of climate change. The radiative balance between
incoming solar shortwave radiation (SWR) and outgoing longwave radiation
(LWR) is influenced by global climate "drivers." Natural fluctuations in solar
output (solar cycles) can cause changes in the energy balance (through fluctuations
in the amount of incoming SWR). Human activity changes the emissions of gases
and aerosols, which are involved in atmospheric chemical reactions, resulting
in modified O_3 (ozone) and aerosol amounts. O_3 and aerosol particles absorb,
scatter, and reflect SWR, changing the energy balance. Some aerosols act as cloud
condensation nuclei, modifying the properties of cloud droplets and possibly
affecting precipitation. Because cloud interactions with SWR and LWR are large,
small changes in the properties of clouds have important implications for the
radiative budget. Anthropogenic changes in greenhouse gases (GHGs, e.g., CO_2,
CH_4, N_2O, O_3, CFCs) and large aerosols (>2.5 μm in size) modify the amount
of outgoing LWR by absorbing outgoing LWR and re-emitting less energy at a
lower temperature. Surface albedo is changed by variations in vegetation or land
surface properties, snow or ice cover, and ocean color. These changes are driven
by natural seasonal and diurnal changes (e.g., snow cover), as well as human
influence (e.g., changes in vegetation types) (Forster et al. 2007). Original source
of figure and legend: reproduced with permission from Stocker et al. (2013: figure
1.1 therein)

Climate evolves over time, and stochastic processes (that is, time-dependent
random variables representing climate variables with not exactly known
values) and time series (that is, the observed or sampled process) are central to

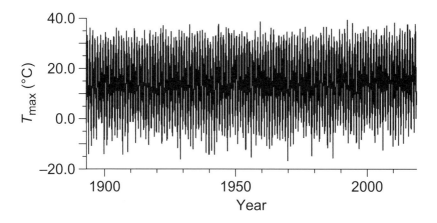

Figure 1.2 Maximum daily surface-air temperature at Potsdam, Germany, during the instrumental period. The interval is from 1 January 1893 to 31 December 2018; the data size is 46,020. The maximum is taken from the three daily temperature readings at 7:00 a.m., 2:00 p.m., and 9:00 p.m. Data from Friedrich-Wilhelm Gerstengarbe (Potsdam Institute for Climate Impact Research, Germany, personal communication, 2014) and www.pik-potsdam.de/services/climate-weather-potsdam/climate-diagrams (6 February 2019).

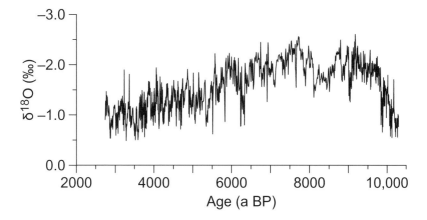

Figure 1.3 Indian Ocean monsoon rainfall during the Holocene. The interval is from 2741 to 10,300 years before the present (BP); the data size is 1345. The climate archive is stalagmite Q5 from Oman, which grew during the shown interval. The proxy variable oxygen isotopic composition (δ^{18}O) indicates the amount of rainfall, with low δ^{18}O reflecting strong monsoon. Time runs from right to left. The vertical scale is inverted in a paleoclimatic manner so that the transition from the last glacial to the present Holocene interglacial at around 10,000 a BP is "upward." Data from Fleitmann et al. (2003)

statistical climate analysis. We use a wide definition of trend and decompose a stochastic process, X, as follows:

$$X(T) = X_{\text{trend}}(T) + X_{\text{ext}}(T) + S(T) \cdot X_{\text{noise}}(T), \qquad (1.1)$$

where T is continuous time, $X_{\text{trend}}(T)$ is the trend component, $X_{\text{ext}}(T)$ is the extreme component, $S(T)$ is a variability function scaling $X_{\text{noise}}(T)$, the noise component. The trend is seen to include all systematic or deterministic, long-term processes, such as a linear increase, a step change, or a seasonal signal. The trend is described by parameters, for example, the rate of an increase. Extremes are events with a large absolute value and are usually rare. The noise process is assumed to be weakly stationary with zero mean and autocorrelation. Giving $X_{\text{noise}}(T)$ standard deviation unity enables the introduction of $S(T)$ to honor climate's definition as not only the mean but also the variability of the state of the atmosphere and other compartments (Brückner 1890). A version of Eq. (1.1) is written for discrete time, $T(i)$, as

$$X(i) = X_{\text{trend}}(i) + X_{\text{ext}}(i) + S(i) \cdot X_{\text{noise}}(i), \qquad (1.2)$$

using the abbreviation $X(i) \equiv X(T(i))$, etc. The observed, discrete time series from process $X(i)$ is the set of size n of paired values $t(i)$ and $x(i)$, compactly written as $\{t(i), x(i)\}_{i=1}^{n}$.

The big question concerns $X_{\text{ext}}(T)$. There are parameters that describe the extreme component. Statistical methods help to estimate the parameters using the time series data.

Box 1.2   Personal Reflection: Statistical notation and the climate equation

I was puzzled, as perhaps you now, when I first saw in statistics books the distinction made between X and x or between T and t. It indeed makes sense to keep the numerical value of a sample (e.g., $x = 20\,°\text{C}$) apart from the probabilistic concept of a random variable (e.g., $X =$ temperature at Potsdam). We do not know the exact value of the current temperature at Potsdam before it is measured. Before the measurement, however, we may be able to draw a curve of the chance of observing a certain temperature, x, against x on the basis of past measurements. This curve is called the probability density function (PDF) of the random variable X. (Strictly speaking, we have to speak of observing a value between x and $x + \delta$, with δ being arbitrarily small.)

The interesting thing with time series is that we now concatenate the values $x(i)$ according to the observation times $t(i)$; the symbol i is the counter. This allows the climate's memory to be taken into account: hot today, likely also hot tomorrow. This memory is mathematically referred to as autocorrelation, persistence, or serial dependence. The concatenated random variables, $X(i)$, are called a stochastic process. Our modeling of the climate process has to take into consideration also trend, extremes, and variability – and then we have the climate equation (Eq. 1.2). Our task is statistical inference: to guess the properties of the climate process, mainly its extreme component, using the sample.

A note on the climate equation: This is a conceptualization that has proven itself to be descriptive of what we know about the course of climate over time. It is not an equation derived from the first principles of physics – this is impossible for the complex, "dirty" climate system (Figure 1.1).

Since the climate variable, $X(T)$, includes the trend, $X_{\text{trend}}(T)$, we have either to assume absent trends or to estimate the trend component and remove it. On top of this, we have to separate $X_{\text{ext}}(T)$ from the noise component, $S(T) \cdot X_{\text{noise}}(T)$, which is also contained in the climate equation. Detection of extremes, their sizes, and when they occurred is the theme of Chapter 2. This chapter also details the different types of extreme data. Appendix A informs about the various climate variables being measured (observations and proxy) and Appendix B informs about climate archives. A special case is the time series that is produced by a climate model (Appendix C).

The statistical language, the methodological core, and the various statistical tools used to analyze the extreme component data are detailed in Chapter 3. These first three chapters put us in a position to apply the tools to real data.

The application chapters focus on the climate variables of strongest influence on society and economy: precipitation, temperature, and wind speed. We will study both positive and negative extremes. From precipitation extremes (Chapter 4), floods refer to positive high values and droughts to small but still positive values. Regarding temperature (Chapter 5), there are heatwaves and cold spells. In the case of wind speed (Chapter 6), we look at hurricanes and other storms but ignore doldrums.

For each of the three variables, precipitation, temperature, and wind speed, many new articles appear each week in the scientific literature, and it is difficult to keep pace. Therefore, we select a few, topical examples of climate extremes, which we analyze as case studies. Each chapter has at the end a section entitled "Reading Material," which contains a selection of major papers or books that have advanced the scientific field.

We hope that the book's structure will provide a solid basis for understanding data and methods. You will be able to apply the described methods to new data and obtain new results, which can then be interpreted. You can test your understanding in the exercises at the end of Chapters 2 and 3. Background information, which may be helpful for doing the exercises, is given on statistical estimation (Appendix D), numerical techniques (Appendix E), and data and software (Appendix F). For course instructors: The solutions to the exercises are given on the book's URL (secured part).

Finally, distributed throughout the book, boxes with "Personal Reflections" appear: stories, personal assessments, or explanations from the author, written in hopefully clear and understandable language. The intention is to make this major theme accessible to the student. The statistical analysis of climate extremes is, as you will experience, a cutting-edge scientific area of high socioeconomic relevance.

Reading Material

Bradley (1999) is a textbook on paleoclimatology, which is structured according to climate archives, dating tools, and proxy variables. Brönnimann et al. (2008) is an edited book on climate during the instrumental period, with a focus on temperature. Cronin (2010) is a textbook on paleoclimatology, which is structured according to the geologic timescale. Peixoto and Oort (1992) is a textbook on climate from a physics viewpoint. Pierrehumbert (2010) is a textbook on climate from a geophysics viewpoint, also on other planets than Earth; see www.cambridge.org/pierrehumbert (6 April 2019). The Fifth Assessment Report of the IPCC (Stocker et al. 2013) gives a comprehensive assessment of climate change; it can be downloaded from www.ipcc.ch (6 April 2019). The Sixth Assessment Report is due in 2021. Brückner (1890) is a long, early paper contributing to climate's definition not only in terms of the mean but also the variability. Fleitmann et al. (2003) is an early paper utilizing the stalagmite climate archive and the $\delta^{18}O$ proxy variable to study past monsoonal rainfall at high temporal resolution.

Bryant (1991) is a textbook on geologic and climate extremes, excluding temperature. Kropp and Schellnhuber (2011) is an edited book on climate extremes and their statistical analysis, with a focus on precipitation and floods. Mudelsee (2014) is a book on the statistical analysis of climate time series; it contains a chapter on extremes; see www.manfredmudelsee.com/book (6 April 2019).

The book by Bell (1986) can be described as mathematics narrated via biographies. Popper (2004) is a talk about realism and accessibility. Russell (1996) is a classic, an original, and it is no masticated food.

2

Data

At the start of the examination, there is the time series, $\{t(i), x(i)\}_{i=1}^{n}$. The task is to infer the properties of the extreme component in the climate equation (Eq. 1.2). This requires the detection and extraction of the extreme values, $\{t_{\text{ext}}(j), x_{\text{ext}}(j)\}_{j=1}^{m}$. These form a subset of the original time series. The size of the subset is less than or equal to the original data size, or, mathematically, $m \leq n$.

Inference is possible if the time values, $\{t_{\text{ext}}(j)\}_{j=1}^{m}$, are known when an extreme occurred (Section 2.1). If the magnitudes, $\{x_{\text{ext}}(j)\}_{j=1}^{m}$, are also of interest, a straightforward method to detect the extremes is to place a threshold for $x(i)$ and analyze the peaks that exceed it (Section 2.2). The alternative is to segment the series into time blocks and take the block maxima (Section 2.3). The presence of a time-dependent trend or variability (Eq. 1.2) may require the employment of robust tools in order to detect extremes (Section 2.4).

"Negative extremes," for example, droughts, can be detected by taking block minima or the peaks below a threshold. The quickest route for such analyses is to multiply the values $x(i)$ beforehand by -1.

2.1 Event Times

The event times are mathematically defined as

$$\{T_{\text{ext}}(j)\}_{j=1}^{m} = \{T(i) \,|\, X_{\text{ext}}(i) \neq 0\}_{i=1}^{n}. \tag{2.1}$$

The symbol "|" means "conditional on." Known are the time points when an extreme occurred and that its magnitude is therefore unequal to zero – the magnitude value itself is unknown or not of interest.

A second constraint imposed on $X_{\text{ext}}(j)$, besides being nonzero, is independence. The observed extreme should have occurred because a climate

10

process generated it and not because there had previously been another, interfering event.

The data type "event times" is used for a variety of archives. For example, a frost ring is, according to Merriam-Webster, a false annual ring in the trunk of a tree that is often evident only as a brownish line. It is caused by defoliation due to frost and the subsequent leafing out again. The accuracy of the magnitude of the cold extreme in this temperature proxy from the tree-ring archive is rather low. It is therefore wise to focus on the event times, $T_{ext}(j)$. Another example is from the documentary archive. Often it is only recorded that an extreme happened at a certain date but not how strong it was.

The event times are the entry data for occurrence rate estimation (Chapter 3).

2.1.1 Elbe Winter Floods

Figure 2.1 shows the flood events of the river Elbe during winter over the past millennium. Hydrological winter is from November to April. The flood

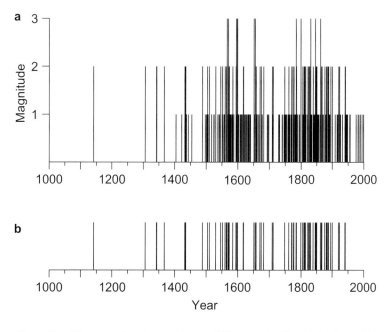

Figure 2.1 Elbe winter floods over the past 1000 years. (a) Magnitude classes (1, minor; 2, strong; 3, exceptionally strong); (b) event times of class 2–3 floods. Data from Mudelsee et al. (2003)

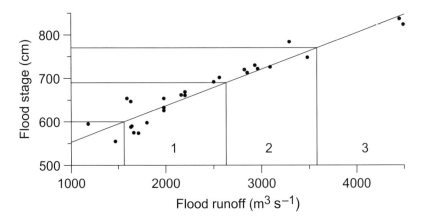

Figure 2.2 Classification of flood magnitudes on the river Elbe. The calibration is obtained by means of a linear regression (solid line). The regression is fitted to data pairs (points) of flood stage (Königliche Elbstrom-Bauverwaltung 1893) and runoff (Global Runoff Data Centre, Koblenz, Germany, personal communication, 2013–2014). The data comprise 27 points, which are from the interval from 1852 to 1891. The magnitude-class boundaries are at 600 cm, 690 cm, 770 cm and 1560 $m^3 \ s^{-1}$, 2630 $m^3 \ s^{-1}$, 3580 $m^3 \ s^{-1}$, respectively. For the method, see Section D.1.2. The magnitude classification corresponds to Figure 2.1a. Adapted from Mudelsee et al. (2003)

magnitude (Figure 2.1a) is divided into three classes. The total number of events is 211. The heavy floods (class 2 or 3) are also shown as event times (Figure 2.1b).

Data for the documentary period before 1850 were extracted from Curt Weikinn's compilation of source texts on hydrography in Europe (Weikinn 1958, 1960, 1961, 1963, 2000, 2002). The accuracy of these flood dates varies; sometimes only the month but not the day is reported. Data for the instrumental period, from 1850, were inferred via a calibration using daily measurements of water stage and runoff (volume per time interval) at the Elbe measurement station Dresden (Figure 2.2). Magnitude classification was possible owing to the existence of sporadic flood-stage measurements during the documentary period.

For the period up to 1850, independence of events was achieved by studying the historical sources (Mudelsee et al. 2003). Consider the ice flood in 1784, for which Weikinn (2000) lists 32 source texts that report the ice cover breaking in the last week of February, the rising water levels, the considerable damages this and the moving ice floes caused, and, finally, the decreasing water levels in the first week of March 1784. This has to be considered as one single event ($t_{\text{ext}}(j) = 1784.167$) and not two (February, March).

The question after the flood risk, whether winter floods occur at a constant rate or whether instead changes exist, is analyzed by means of occurrence rate estimation (Chapter 3).

Box 2.1   Personal Reflection: Weikinn and the Weikinn source texts

It was my privilege to work with the Weikinn source texts of documentary climate data and their editors during the early 2000s. I came from a postdoctoral stay at the Institute of Mathematics and Statistics at the University of Kent in the United Kingdom. There I had developed a concept to implement nonstationary risk analysis tools (Chapter 3), which I wanted to apply in practice. The Weikinn texts on hydrography covering the interval from AD 0 (in the language of the German Democratic Republic called "Zeitwende") to 1750 had already been published (Weikinn 1958, 1960, 1961, 1963). Michael Börngen and Gerd Tetzlaff were working at the Institute of Meteorology of the University of Leipzig in Germany on the edition for the interval from 1750 to 1850.

The editorial work included collecting and systematizing the vast amount of entries (23,160 for the full time interval), handwritten by Weikinn, and putting them into a digital (i.e., searchable) form. For the climate extremes analysis, we decided to analyze the floods of the rivers Elbe and Oder. The English translation of the German word "oder" is "or," so this work included an extra challenge when interpreting Weikinn's notes. (It would have been too risky to rest only on the separability based on the capitalization of these words.)

The biggest danger with documentary climate data are inhomogeneities via document loss. There are fewer documents from earlier periods about extreme events as these were before the invention of printing (in Europe around 1450). Some earlier documentation, however, is available to us through secondary sources, such as Weikinn's compilation. Still, this can lead to missed events and an underestimation of the flood occurrence rate (Chapter 3). Also, the perception of extreme events such as floods and the willingness of people to record weather events might have been different before the Copernican scientific revolution.

Curt Weikinn (1888–1966) was an autodidact, who earned his living in the German Democratic Republic as a bank clerk. He was diligent and had access to a vast amount of literature – original documents and secondary sources. After his death he left source texts not only about hydrography but about weather conditions in general, most of which are still unpublished. The fact that he had no academic affiliation, and that he relied heavily on secondary sources, has brought his source texts under criticism. Also our manuscript (Mudelsee et al. 2003) received critical review comments in this direction when it was submitted to the high-impact journal *Nature*.

For our resubmission, we therefore produced additional flood records on the basis of the compilation by the historian Militzer (1998), which covers the interval from 1500 to 1799. It turned out that the estimated time-dependent flood occurrence rates in the overlapping time interval were, within the error band, indistinguishable from each other (Chapter 4). Two main conclusions may be drawn. First, it is advisable for a climatologist to collaborate with historians trained in interpreting original sources. Second, it can be said that Weikinn did a good job.

2.2 Peaks Over Threshold

The peaks over threshold (POT) extremes are defined as

$$\left\{T_{\text{ext}}(j), X'_{\text{ext}}(j)\right\}_{j=1}^{m} = \left\{T(i), X(i) \,\middle|\, X(i) > u\right\}_{i=1}^{n}. \tag{2.2}$$

This definition thus requires that the variable, $X(i)$, is above a constant threshold, u. Not only the event times, $T_{\text{ext}}(j)$, but also the values of the variable, $X'_{\text{ext}}(j)$, are taken. (The prime, written for reasons of convention, is later dropped.)

The POT approach is useful for data whose magnitude is known with a reasonable accuracy, and not just the date. In climate sciences it is also useful to consider a time-dependent threshold to take into account effects of trends in mean, $X_{\text{trend}}(i)$, and variability, $S(i)$. To satisfy the assumption of mutual independence of the POT data, imposing further criteria rather than passing the threshold may be necessary. This is pursued in Section 2.4.

The assumption that the values of $X'_{\text{ext}}(j)$ are mutually independent, is made also for the POT extremes. These are the entry data for fitting a Generalized Pareto (GP) distribution and estimating the occurrence rate (Chapter 3).

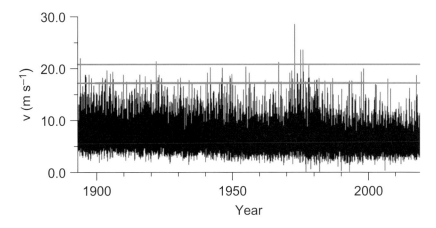

Figure 2.3 Threshold data, maximum daily wind speed at Potsdam. The interval is from 1 January 1893 to 31 December 2018, the data size is $n = 46{,}020$. Two thresholds ($u = 17.2$ m s^{-1} and 20.8 m s^{-1} are shown (gray horizontal lines). The number of events equal to or greater than u, is $m = 7$ for the larger threshold and $m = 93$ for the smaller threshold. The daily maximum is taken from the 24 hourly means. Data from www.pik-potsdam.de/services/climate-weather-potsdam/climate-diagrams (8 May 2019), with two missing values (31 May 2017 and 31 May 2018) inserted from ftp://ftp-cdc.dwd.de/pub/CDC/observations_germany/climate/hourly/wind/historical (8 May 2019)

2.2.1 Potsdam Wind Speed

Figure 2.3 shows the time series of maximum daily wind speed at Potsdam, Germany. No distinction is made into seasons.

The Potsdam series has an important property: homogeneity. This means that neither the measurement setting nor the observation times have changed, at least over the shown interval since 1893 (Körber 1993). Homogeneous records allow a more straightforward analysis and interpretation. Inhomogeneous records may require data preprocessing – which introduces new uncertainties.

There are two upper thresholds, u, shown in Figure 2.3. The larger one, at 9 Beaufort or 20.8 m s^{-1}, defines meteorologically a storm. The smaller one, at 8 Beaufort or 17.2 m s^{-1}, is the level relevant for the insurance industry in Germany. It is the convention in this sector to define as extreme event the case "$\geq u$" and not just "$> u$."

2.3 Block Extremes

The annual maximum is a common example of a block maximum. The formula is

$$X'_{\text{ext}}(j) = \max\Big(\big\{X(i)\big\}|T(i) \text{ within } j\text{th year of time series}\Big), \qquad (2.3)$$

$$T_{\text{ext}}(j) = j\text{th year of time series.} \qquad (2.4)$$

This extremes data type is illustrated using the Potsdam wind-speed data (Figure 2.4).

The block extremes, $X'_{\text{ext}}(j)$, are the input for fitting a Generalized Extreme Value (GEV) distribution (Section 3), which can shed light on the risk that an extreme of a predefined size and at a predefined block length occurs.

Risk estimation assumes that an extreme is taken from a block with a large number, k, of independent observations (at least, say, 100). This can be done explicitly by segmenting or "blocking" an original series, $\{X(i)\}_{i=1}^{n}$. Alternatively, the blocking may have already been done implicitly; for example, documentary data in the form of maximum annual water stage in a river, where the daily observations have been deleted to free computer storage space. The persistence time (Eq. 2.8) measures how long a climate series memorizes. If the block length is large compared with the persistence time, then the

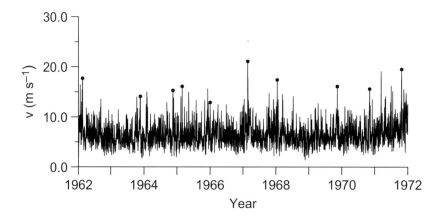

Figure 2.4 Block maxima, maximum daily wind speed at Potsdam. The series (solid line) is plotted for the interval from 1 January 1962 to 31 December 1971. For the full series, see Figure 2.3. The selected block length is one year. The total number of annual maxima (filled symbols) over the full interval from 1893 to 2018 is $m = 126$.

assumption that the values of $X'_{\text{ext}}(j)$ are mutually independent, should be approximately fulfilled.

2.4 Detection of Extremes

When attempting to detect extremes, obstacles can arise in practice. In the POT approach, the challenge is the placement of the threshold (Section 2.4.1). In the blocking approach, block length selection may also be difficult (Section 2.4.2).

2.4.1 Threshold Placement

A climate variable comprises trend, extreme, and noise components (Eq. 1.2). The threshold approach detects the extremes against trend and noise. For time-stationary trend and noise components, it is possible to set a constant threshold (as in Figure 2.3). For time-dependent trend, $X_{\text{trend}}(i)$, and noise variability, $S(i)$, it may be useful to replace the time-constant threshold, u, by a time-dependent function. It is mandatory to estimate $X_{\text{trend}}(i)$ and $S(i)$ in a robust manner, that means, without interference by the extreme component, $X_{\text{ext}}(i)$.

A suitable, robust tool for $X_{\text{trend}}(i)$ estimation is the running median. The median is calculated from $2\,l + 1$ points inside a pointwise shifted window. Likewise, the running median of absolute distances to the median (MAD) is a robust $S(i)$ estimator. The time-dependent threshold is then given by median $+ z \cdot$ MAD. The threshold parameter, z, has to be selected.

In the example (Figure 2.5), the topmost point inside the window does not affect trend or variability estimation. We would obtain the same result if that point were farther away from the other points by a factor of, say, 100. On the other hand, the running mean and standard deviation are not robust measures of trend and variability, respectively.

Figure 2.6 shows detection of extremes in the varve thickness record from the Lower Mystic Lake in the Boston area. The background trend variations represent a combination of natural and anthropogenic (colonization) factors.

The character of the laminated sedimentation changed after around 1870. This was the result of population growth, industrialization in the watershed, and permanent alteration of the lake's natural hydraulic regime due to dam building (Besonen et al. 2008). To prevent the influence of these inhomogeneity factors on the detection of peaks in varve thickness, the twentieth-century part of the lake core is not considered. The task is to detect the peaks in varve

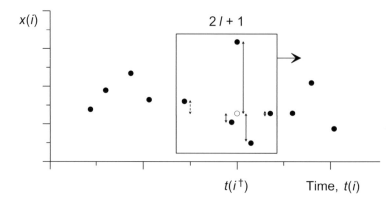

Figure 2.5 Running median and MAD. A window containing $2\,l + 1$ points is used; here, $l = 2$. The analyzed time point, $t(i^\dagger)$, is the median of the window time points. The trend at $t(i^\dagger)$ is estimated as the median, that is, the $(l + 1)$ largest, of the window data points (open circle). The variability at $t(i^\dagger)$ is estimated as the MAD (dotted double arrow). Also shown (solid double arrows) are the absolute distances to the median for the other window points; the rightmost double arrow has a length of zero. For estimating trend and variability over the whole series, the window is shifted pointwise. When approaching the interval bounds, $i \to 1$ or $i \to n$, a simple solution is to extrapolate trend and variability there.

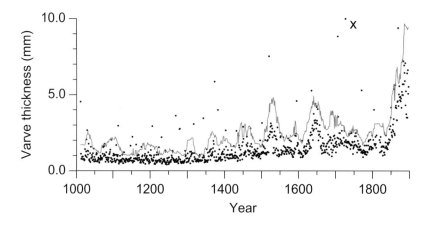

Figure 2.6 Extremes detection in the Lower Mystic Lake varve thickness record. Shown are the annual varve thickness (filled symbols) and the detection threshold (gray line). Detection parameters: $l = 8, z = 5.2$. (One data point (cross) with true $x = 16.7$ mm is shown here in this plot at 10.0 mm for better legibility.) Data from Besonen et al. (2008)

thickness for the shown interval, which are interpreted to have arisen from hurricanes that moved through the site region.

In addition to varve thickness, Besonen et al. (2008) determined the dates of graded beds. This, together with a thick varve, can jointly be used as a proxy for hurricane activity in the vicinity of the lake. Hurricane-strength precipitation saturates the watershed, resulting in erosive overland flow that entrains sediment and carries it into the lake, where it is deposited as a graded bed. This effect is enhanced by hurricane-strength winds that disturb vegetation and uproot trees, exposing loose sediment. The proxy information was verified by means of pollen data and documentary information (available from about 1630 to the present).

A number of 17 window points were selected for calculating the running median and MAD. This means that trend and variability variations on decadal timescales are permitted. Several threshold selections were evaluated and the optimal compromise (liberal versus conservative) can be seen in $z = 5.2 \approx 3.5/0.6745$ (Besonen et al. 2008). This corresponds to 3.5 robust standard deviations. (A normal distribution with standard deviation unity has an MAD of ~ 0.6745.)

The number of peaks detected in this manner is 47. The second condition imposed to a varve is the existence of a graded bed. This led Besonen et al. (2008) to discard 11 of those events. The further analysis of the hurricane activity (Chapter 3) is therefore based on 36 events, observed between 1011 and 1897.

2.4.2 Block Length Selection

To understand the concept of block maxima, consider one random variable, X, with a certain PDF, $f(x)$. For example, the PDF may be Gaussian. We ask what is the chance of observing X with a value greater than or equal to a value, u. The answer is given by the area under the PDF, the integral, $\int_u^\infty f(x)\,dx$. Next, consider two random variables, X_1 and X_2, both with the same PDF. Assume independence, that means, values drawn from X_1 do not depend on the values drawn from X_2, and vice versa. We now ask what is the chance of observing the block maximum, $\max(X_1, X_2)$, with a value greater than or equal to u. It is possible to derive an analytical solution for the PDF of the maximum, but that is not the point here. Let us consider k random variables, $X_1, \ldots, X_k$, all with the same PDF. The point is that with increasing k, the block maximum follows approximately a distribution – the GEV – irrespective of what shape the individual variables have. This is illustrated in Figure 2.7.

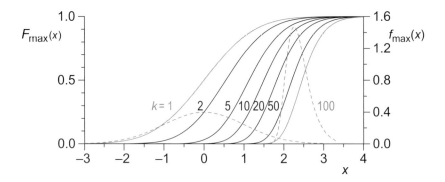

Figure 2.7 Distribution of the maximum of k independent standard normal variates. The distribution functions, $F_{\max}(x)$ (solid lines), are labelled with their k value. For $k = 1$ and 100 (both gray color), the density functions, $f_{\max}(x)$, are also shown (dashed lines). Letting k increase has three effects: the location (average) is shifted to the right, the scale (standard deviation) is decreased, and the right-skewness (shape parameter) is increased. With increasing k, the distribution of the maximum approaches the GEV shape.

Box 2.2   Personal Reflection: Gauß and the Gaussian distribution

In many settings of measurements in the natural sciences, the random effects add to each other, such as in concentration measurements in chemistry. Irrespective of which distributional shape an individual random fluctuation follows, the total or average effect follows approximately a normal distribution, which is also called a Gaussian distribution. This central limit theorem is the root of the importance and ubiquity of the Gaussian distribution.

The Gaussian PDF is given by

$$f(x) = (2\pi)^{-1/2} \exp\left(-0.5(x - \mu)^2\sigma^{-2}\right). \qquad (2.5)$$

If the mean, μ, equals zero and the standard deviation, σ, equals unity, then the PDF is called a standard normal distribution (Figure 2.7). The Gaussian distribution has a median of μ and an MAD of $\sim$0.6745 σ.

The approximation by the Gaussian distribution is better, that means, closer to the limit, the more random effects contribute. Carl Friedrich Gauß (1777–1855) was instrumental in sharpening the definition and proving the existence of a limit value in mathematics. His collected works, bound in light brown suede, inspired awe in a young student sitting in the library of the Institute of Applied Mathematics at the University of Heidelberg in Germany. What stayed in my mind from a cursory occupation with this man was that he is considered by Bell (1986) to be one of the three giants in mathematics. His writing style produced concise, polished works in Latin, but without traces of how he had proceeded. He gave Georg Friedrich Bernhard Riemann (1826–1866) a headache in the preparation of his habilitation defense, which paved the way for measuring space more generally (non-Euclidean geometry). By the way, the normal or Gaussian distribution had actually been found earlier by Abraham de Moivre (1667–1754).

How large should the number, k, of block elements be to make the GEV approximation accurate? Consider the example of daily data. Annual maxima ($k = 365$ or 366) should yield a good result while monthly maxima would less so. Evidently, the data size, n, should influence the selection of k.

What if there is dependence among the variables, $X_1, \ldots, X_k$? Such a situation is indeed typical for climate with its memory (Chapter 1). Assume for heuristic purposes a dependence of the simple form, $X_1 = X_2, X_3 = X_4$, and so forth. Then the number of independent block elements is $k/2$. Persistence, irrespective of its mathematical model, reduzes the size of independent data.

A suitable model of climate's persistence is the first-order autoregressive or AR(1) process,

$$
\begin{aligned}
X_{\text{noise}}(1) &= \mathcal{E}_{\text{N}(0,\,1)}(1), \\
X_{\text{noise}}(i) &= a \cdot X_{\text{noise}}(i - 1) + \mathcal{E}_{\text{N}(0,\,1-a^2)}(i), \qquad i = 2, \ldots, n.
\end{aligned}
\tag{2.6}
$$

Herein, $-1 < a < 1$ is a parameter and $\mathcal{E}_{\text{N}(\mu,\,\sigma^2)}(i)$ is a Gaussian purely random process ("innovation") with mean μ, variance σ^2, and no dependence. The larger a, the stronger the memory and the smaller the innovation; $a = 0$ means a purely random process; $a < 0$ is rarely found in climate sciences. This AR(1) process has time-constant mean zero and variance unity, and, as a result, is called stationary. (Strictly speaking, these two properties, mean and variance, define weak stationarity. If also higher-order properties, such as skewness, are time-constant, then the process is called strictly stationary. A weakly stationary process with a Gaussian shape is also a strictly stationary process.)

An estimator of the autocorrelation parameter, that means, a recipe how to calculate a from noise data, $\{x_{\text{noise}}(i)\}_{i=1}^{n}$, is given by

$$
\widehat{a} = \sum_{i=2}^{n} x_{\text{noise}}(i) \cdot x_{\text{noise}}(i - 1) \left/ \sum_{i=2}^{n} x_{\text{noise}}(i)^2 \right. .
\tag{2.7}
$$

(See Appendix D for estimators and hat notation.) Estimator $\hat{a}$ is biased, that means, if $\{X_{\text{noise}}(i)\}$ is an AR(1) process with parameter a, then its expectation, $E(\hat{a})$, is unequal to a. Only approximation formulas exist for the bias in the autocorrelation estimation. Such formulas can be used for bias correction.

The AR(1) process written as Eq. (2.6) and the estimator (Eq. 2.7) of the autocorrelation parameter, a, apply to the case of even time spacing. Climate records obtained from natural archives, however, often show uneven spacing (Appendix B). The AR(1) process for uneven spacing is written as

$$
\begin{aligned}
X_{\text{noise}}(1) &= \mathcal{E}_{\text{N}(0,\ 1)}(1), \\
X_{\text{noise}}(i) &= \exp\left\{-\left[T(i) - T(i-1)\right]/\tau\right\} \cdot X_{\text{noise}}(i-1) \\
&\quad + \mathcal{E}_{\text{N}(0,\ 1-\exp\{-2[T(i)-T(i-1)]/\tau\})}(i), \qquad i = 2, \ldots, n.
\end{aligned}
\tag{2.8}
$$

Herein, the parameter $\tau > 0$ is the persistence time (Mudelsee 2014: chapter 2 therein). The estimation of τ from noise time series, $\{t(i), x_{\text{noise}}(i)\}_{i=1}^{n}$, can be achieved using a least-squares criterion (Section D.1), but the calculation requires numerical techniques (Section E.1). The model (Eq. 2.8) includes also the case of even spacing, where $t(i) - t(i-1) = d$ is constant. Then the persistence parameters are related as $a = \exp(-d/\tau)$.

Exercises

Exercise 2.1 Elbe Flood Dates

Locate on the internet the supplementary data file for Mudelsee et al. (2003) with the Elbe flood catalogue. Produce a data file of type "event times" for the heavy winter floods (magnitude classes 2 or 3). Hints: Use a software that allows text extraction from PDF files. Transform the flood dates into real time values. A plot of the data is shown in Figure 2.1b. The data file should be a text file with one column that contains $t_{\text{ext}}(j)$ in increasing order. The values should have neither a thousand comma nor a decimal comma (but a decimal point is allowed).

Exercise 2.2 Homogeneity

Consider a rural meteorological station in the year 1893. The regional temperature at the site is assumed not to show a long-term trend. The temperature is measured each day at 7:00 a.m., 1:00 p.m., and 6:00 p.m. Mean and maximum temperatures per day are calculated from the three measurements as the mean and the maximum, respectively. From the year 1946, the measurement time of the morning temperature switched to 8:00 a.m.

How does the switch in measurement time influence:

1. mean temperature;
2. maximum temperature?

What effects do you anticipate in the following settings:

3. a rural station and the presence of a regional warming trend;
4. a rural station and the presence of a regional cooling trend;
5. an urban setting, with no switch in measurement time and no regional trend?

Exercise 2.3 Detection of POT Extremes

This is a software exercise using the risk analysis program Caliza. See Section F.2 for access to the software and its manual. Read the manual up to Part 2.

Start the software. Select the varve thickness record from the Lower Mystic Lake (Figure 2.6), which is available as file LML.txt (Section F.1).

Part I of the software, time interval extraction, is to be ignored here.

Part II of the software is the detection of POT extremes against a time-dependent threshold. This task involves basically setting two parameters. First, the number of points contained in the running window (Figure 2.5) is equal to $2\,l + 1$. Second, the threshold factor is given by the parameter z (Figure 2.6).

Regarding the selection of l, note that a large value means more window points and, hence, a more precise estimation of the median and the MAD (Figure 2.5). A drawback of a large l value is that real, systematic trend or variability features are smoothed away, introducing bias. On the other hand, a small l value generates less precise estimations and reduces bias. This problem, between precision and bias, is called the smoothing problem in statistical science. The solution lies somewhere in the middle.

Cross-validation is one approach to solve the smoothing problem. Its concept is based on optimally predicting the trend for a certain time point by utilizing the window points. This is done for all time points. The optimal l optimizes the average prediction. That means, it minimizes a function, $C(l)$, of l. Caliza consults two cross-validation functions,

$$ C_1(l) = \left[\sum_{i=1}^{n} \left| x(i) - \widehat{m}\,\{x(j)\}_{j=i-l,\ j\neq i}^{i+l} \right| \right] \Big/ n, \qquad (2.9) $$

$$C_m(l) = \widehat{m} \left\{ \left| x(i) - \widehat{m} \{x(j)\}_{j=i-l,\ j \neq i}^{i+l} \right| \right\}_{i=1}^{n}. \tag{2.10}$$

Within these equations, the inner median, $\widehat{m} \{x(j)\}_{j=i-l,\ j \neq i}^{i+l}$, is the delete-one background estimate. The median sits in the middle of the data; half of the data are larger and half are smaller. One leaves out the point $j = i$ to exclude the trivial solution $l = 0$. However, because those criteria assume absent serial dependence, it is important in a practical application to try different l values and study the sensitivity.

Regarding the selection of z, note that a large value means a higher threshold and potentially fewer detected POT extremes. This propagates into larger uncertainties in the risk estimation (Chapter 3). On the other hand, a smaller z value means more extremes and more accurate risk estimations. However, too small z values weaken the POT criterion. A good choice between too high (conservative approach) and too small (liberal approach) is in the middle. The selection of z constitutes a statistical trade-off problem. Again, it is important in a practical application to try different z values and study the sensitivity.

The exercise consists in playing with various combinations of l and z and noting the effects on the threshold and the number of detected extremes. Reproduce the setting from Figure 2.6, with $l = 8$ and $z = 5.2$, leading to $m = 47$ POT extremes. What is the average window length in time units?

Exercise 2.4 Stationarity

For a random variable, X, with PDF, $f(x)$, the mean, μ, and the variance, σ^2, is defined by

$$\mu = \int_{-\infty}^{+\infty} x\, f(x)\, dx, \tag{2.11}$$

$$\sigma^2 = \int_{-\infty}^{+\infty} (x - \mu)^2\, f(x)\, dx, \tag{2.12}$$

respectively. Loosely speaking, μ describes the central value and σ^2 describes the spread of the distribution around the central value. The standard deviation, σ, is a handy quantity since it has the same units as the data.

The expectation operator, E, for a function, $g(X)$, of X, is defined by

$$E\left[g(X)\right] = \int_{-\infty}^{+\infty} g(x)\, f(x)\, dx. \qquad (2.13)$$

Hence, if $g(X) = X$, then $E[X] = \mu$.

The variance operator, VAR, for a function, $g(X)$, is defined by

$$VAR\left[g(X)\right] = \int_{-\infty}^{+\infty} (g(x) - \mu)^2\, f(x)\, dx. \qquad (2.14)$$

Hence, if $g(X) = X$, then $VAR[X] = \sigma^2$.

Thus, if $g(X)$ is a linear function, $g(X) = b \cdot X + c$, where b and c are parameters, then

$$E\left[b \cdot X + c\right] = b \cdot E\left[X\right] + c, \qquad (2.15)$$
$$VAR\left[b \cdot X + c\right] = b^2 \cdot VAR\left[X\right]. \qquad (2.16)$$

Show that the AR(1) process (Eq. 2.6) is weakly stationary, that means, it has time-constant mean and time-constant variance.

Exercise 2.5 Bias Correction

Note that noise data $\{x_{\text{noise}}(i)\}_{i=1}^{n}$ are obtained using Eq. (1.2) by subtracting trend and extremes, and then dividing by the variability. In the simple case of a constant mean, this corresponds to the autocorrelation estimator,

$$\widehat{a} = \sum_{i=2}^{n} [x_{\text{noise}}(i) - \bar{x}_{\text{noise}}] \cdot [x_{\text{noise}}(i-1) - \bar{x}_{\text{noise}}] \Big/ \sum_{i=2}^{n} [x_{\text{noise}}(i) - \bar{x}_{\text{noise}}]^2, \qquad (2.17)$$

where $\bar{x}_{\text{noise}} = \sum_{i=1}^{n} x_{\text{noise}}(i)/n$ is the sample mean. For this estimator, Kendall (1954) determined the approximate expectation,

$$E\left(\widehat{a}\right) \simeq a - (1 + 3a)/(n-1). \qquad (2.18)$$

Use this formula to calculate a bias correction for $\widehat{a}$.

Reading Material

Besonen et al. (2008) is a paper that extends the hurricane database for the Boston area back in time by means of a varve thickness record from a lake sediment core. Königliche Elbstrom-Bauverwaltung (1893) is a report that contains water stages of the Elbe for some floods during the nineteenth century. Körber (1993) describes the history of the meteorological observatory Potsdam. Mudelsee et al. (2003) is a paper that contains flood catalogues for the rivers Elbe and Oder over the past millennium. Militzer (1998) is a collection of historically critically interpreted documentary climate source texts. Weikinn (1958, 1960, 1961, 1963, 2000, 2002) form the Weikinn source texts of documentary climate data.

Box (1953) is a classic paper, which introduces the adjective "robust" for methods that work well even if model assumptions are violated. Johnson et al. (1994) is part of a series of books that provide exhaustive information on statistical distributions; this book includes the Gaussian distribution. Tukey (1977) is a classic textbook on robust statistical methods.

3

Methods

The application of statistical methods to the data (Chapter 2) allows us to make an estimation of the extreme component of the climate (Chapter 1). There exists a well-elaborated mathematical theory and an estimation machinery, which goes back to the 1920s, for the situation of a stationary extreme component (Section 3.1), for which the properties are constant over time. The two stationary models are the GEV distribution for block extremes and the GP distribution for POT data.

In the context of climate change, however, it is more relevant to consider nonstationary models (Section 3.2), for which the properties may change over time. It seems straightforward to extend the stationary GEV and GP models by means of a time dependence in the parameters. However, there are two problems with this extension. First, fitting such models to data becomes a formidable optimization problem. Second, there is a principal limitation due to the restricted functional form of the time dependence. Therefore, the nonstationary or inhomogeneous Poisson process offers more flexibility. Also this method is theoretically understood and supported by powerful estimation procedures, which were developed not in the 1920s but the 1990s.

3.1 Stationary Processes

Stationarity means that the properties of distribution and persistence are constant over time.

3.1.1 GEV Distribution

The GEV distribution is used for data in the form of block maxima (Section 2.3). It has the following distribution function:

$$F_{\mathrm{GEV}}(x_{\mathrm{ext}}) = \begin{cases} \exp\left\{-\left[1 + \xi\,(x_{\mathrm{ext}} - \mu)/\sigma\right]^{-1/\xi}\right\} & (\xi \neq 0), \\ \exp\left\{-\exp\left[-(x_{\mathrm{ext}} - \mu)/\sigma\right]\right\} & (\xi = 0), \end{cases} \tag{3.1}$$

where $1 + \xi\,(x_{\mathrm{ext}} - \mu)/\sigma > 0$, $-\infty < \mu < \infty$, $\sigma > 0$, and $-\infty < \xi < \infty$. The parameters μ and σ identify location and scale of the distribution, respectively. The shape parameter, ξ, determines the behavior in the tails, that is, the extremal part, of the distribution. Plots of the GEV density function are shown in Figure 3.1.

Given a sample of size m of block maxima data, $\{x_{\mathrm{ext}}(j)\}_{j=1}^{m}$, we wish to know the GEV parameters μ, σ, and ξ. The standard method for performing this statistical inference task is maximum likelihood estimation. This requires to maximize the log-likelihood function, $l(\mu, \sigma, \xi)$, which is given by

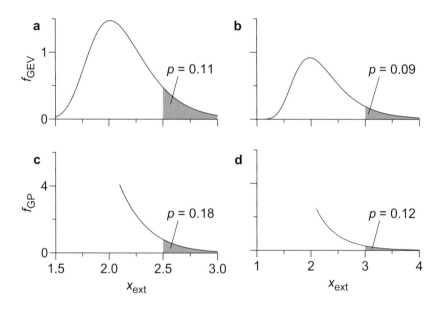

Figure 3.1 GEV and GP density functions. The parameter setting is (a) $\mu = 2.0$, $\sigma = 0.25$, $\xi = -0.05$; (b) $\mu = 2.0$, $\sigma = 0.4$, $\xi = 0.05$; (c) $\mu = 2.0$, $\sigma = 0.25$, $\xi = -0.05$, $u = 2.1$; and (d) $\mu = 2.0$, $\sigma = 0.25$, $\xi = 0.05$, $u = 2.1$. Also shown (shaded areas) are the tail probabilities, p, for two return levels: (a, c) $x_p = 2.5$ and (b, d) $x_p = 3.0$.

$$l\left(\mu,\sigma,\xi\right) = -m\ln\left(\sigma\right) - \left(1 + 1/\xi\right)\sum_{j=1}^{m}\ln\left[y(j)\right] - \sum_{j=1}^{m}y(j), \qquad (3.2)$$

where

$$y(j) = 1 + \xi\left[\frac{x_{ext}(j) - \mu}{\sigma}\right]. \qquad (3.3)$$

That means, the task is to find the parameter values for which the log-likelihood function has its maximum. Numerical techniques (Section E.1) have to be employed for achieving this. The beautiful, concise mathematical notation for the estimation procedure is

$$\left(\widehat{\mu},\widehat{\sigma},\widehat{\xi}\right) = \arg\max\left[l(\mu,\sigma,\xi)\right]. \qquad (3.4)$$

The hat reminds us that we are performing an estimation: $\widehat{\mu}$ does not exactly equal μ. The typical size of this deviation is reported in the form of a statistical uncertainty measure, such as a standard error (Appendix D).

The idea behind maximum likelihood estimation is explained in Figure 3.3 and Box 3.1.

We have implicitly made some assumptions and employed some constraints for the maximum likelihood estimation of the GEV parameters. Sufficiently many independent elements per block should be available (Section 2.4.2). The constraints to the parameter ranges are noted directly after Eq. (3.1). There are further so-called regularity conditions (Coles 2001), on which we remark that the condition $\xi > -0.5$ is usually fulfilled in practical applications, which means that maximum likelihood estimation poses no major technical problems.

Under these conditions, the standard errors of the estimated GEV parameters can also be calculated. The calculations go via the second derivative of the likelihood function, L, which is given via $l(\mu,\sigma,\xi) = \ln(L(\mu,\sigma,\xi))$. To conclude, there exists a well-elaborated mathematical theory and estimation machinery for fitting a stationary GEV distribution to block extremes by means of maximum likelihood.

3.1.2 GP Distribution

GP distribution is used for POT data (Section 2.2). It has the following distribution function:

$$F_{GP}(x_{ext}) = \begin{cases} 1 - \left\{1 + \xi\left(x_{ext} - u\right)/\left[\sigma + \xi\left(u - \mu\right)\right]\right\}^{-1/\xi} & (\xi \neq 0), \\ 1 - \exp\left[-\left(x_{ext} - u\right)/\sigma\right] & (\xi = 0), \end{cases}$$
$$(3.5)$$

where $x_{ext} > u$, $\sigma > 0$, $\left\{ 1 + \xi \left(x_{ext} - u \right) / \left[\sigma + \xi \left(u - \mu \right) \right] \right\} > 0$, and $-\infty <$ $\xi < \infty$. As the GEV, the GP distribution is also described by three parameters, and here also the parameter ξ determines the tail behavior. Plots of the GP density function are shown in Figure 3.1.

Given a sample of size m of POT extremes, $\{x_{ext}(j)\}_{j=1}^{m}$, obtained by means of placement of a threshold, u, the maximum likelihood technique can be used to estimate the GP parameters μ, σ, and ξ. The log-likelihood function to be maximized is

$$l(\tilde{\sigma}, \xi) = -m \ln(\tilde{\sigma}) - (1 + 1/\xi) \sum_{j=1}^{m} \ln[y(j)], \qquad (3.6)$$

where

$$y(j) = 1 + \xi \left[\frac{x_{ext}(j) - u}{\tilde{\sigma}} \right] \qquad (3.7)$$

and

$$\tilde{\sigma} = \sigma + \xi(u - \mu). \qquad (3.8)$$

The constraints to the GP parameter ranges are noted directly after Eq. (3.5). Further conditions are that the threshold is not too small (compared with the center of location of the distribution of the original data) and that $y(j) > 0 \,\forall\, j$. (The symbol $\forall$ means "for all.") As for the GEV, the standard errors for the GP parameters can also be derived from the second derivative of the likelihood function.

3.1.3 Return Level, Return Period, and Risk

Let us consider a time series, $\{t(i), x(i)\}_{i=1}^{n}$, and the set of extreme values, $\{x_{ext}(j)\}_{j=1}^{m}$, which have been detected in the time series by means of blocking or placement of a threshold. We now ask the question after the risk: What is the probability, p, of observing an extreme value that exceeds a predefined, large value, x_p? The answer is obtained by consultation of the distributions of the extremes (Figure 3.1), either the GEV or the GP form. That tail probability, p, is given by the area under the density function. For example, $p = \int_{x_p}^{\infty} f_{GEV}(x_{ext}) \, dx_{ext} = 1 - F_{GEV}(x_p)$ in the case of the GEV form. The value x_p is called return level.

Related to the return level is the return period, which is approximately the expected time span required for observing an extreme value that exceeds the return level. The return period has the numerical value $1/p$; it has the same units as the time values, $t(i)$. Figure 3.1 displays some cases. Let us assume that the time unit, from which the GEV density function in Figure 3.1a has been obtained, is years. The tail probability corresponding to the return level $x_p = 2.5$ equals $p = 0.11$, and an event where x_p is exceeded is called a nine-year event.

The crucial parameter to determine on the basis of data is p, the tail probability. We follow statistical convention and denote p as "risk." Insurance

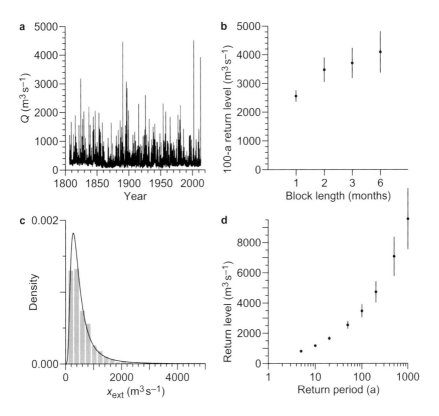

Figure 3.2 Elbe summer floods, GEV estimation. (a) Daily runoff, $x(i)$, at station Dresden for the hydrological summer (May to October) for the interval from May 1806 to October 2013 ($n = 38{,}272$). Data from Global Runoff Data Centre (Koblenz, Germany, personal communication, 2013–2014). (b) 100-year return level, x_p, where $p = 0.01$, in dependence on block length. In hydrology, this x_p level is also denoted as HQ_{100}. (c) Estimated GEV density function (solid line) and histogram estimate of empirical density for block length two months. (d) Return level for fitted GEV model (block length two months) in dependence on return period. The maximum likelihood technique was used for estimation of GEV parameters and their standard errors; the standard errors for the return level (b, d; vertical bars) resulted from error propagation.

services depend heavily on accurate knowledge of the risk. For example, the expected losses (in monetary units) result from multiplying the insured values (in monetary units) by the risk value. Be aware that many definitions of the term "risk" are in usage. Another thing to note is that block extremes (the basis for the GEV distribution) are not the same as POT data (the basis for the GP distribution). This has to be taken into account for risk assessments, where we speak, for example, of "the risk of observing an annual maximum in excess of . . ."

Consider the Elbe summer floods (Figure 3.2) as an illustration of the concept. The daily runoff time series from Elbe station Dresden (Figure 3.2a) is one of the longest available continuous records (no missing data); it should lead to an accurate estimation. The 100-year return level (Figure 3.2b) can be calculated from the estimated GEV parameters (Exercise 3.1). A block length of two months seems adequate, for which the return level approaches within error bars a saturation level. Also the fitted GEV density shows good agreement with a histogram for this block length (Figure 3.2c). Finally, note the increases of the return level and its standard error with the return period (Figure 3.2d). See Exercise 3.2 for replicating this analysis of the Elbe summer floods.

Box 3.1   **Personal Reflection: Fisher and maximum likelihood**

We climatologists work with noisy data generated by the complex climate system. Inspired by the data, we ask quantitative questions and try to solve them by means of statistical inference (Appendix D). Maximum likelihood is a widely used, powerful inference technique. Quasi-automatically, it produces not only estimates but also the associated uncertainties. In addition, the uncertainties (bias and standard error) are usually at least as good (i.e., small) as what competing estimators (e.g., least squares) achieve.

To appreciate the idea behind maximum likelihood, let us consider the data (e.g., the detected extremes, which form a sample of size m) and a parametric model for the data generating process (e.g., a GEV model). For notational convenience, we write the data as an m-dimensional vector, x. We further combine the three GEV parameters into a vector, $\theta = (\mu, \sigma, \xi)'$. The task is then to search for the parameter estimate, $\widehat{\theta}$, for which the PDF of the observed data, x_{obs}, has a maximum (Figure 3.3). Usually, it is mathematically more convenient to maximize the log-likelihood function, $l(\theta) = \ln(L(\theta))$. The idea behind maximum

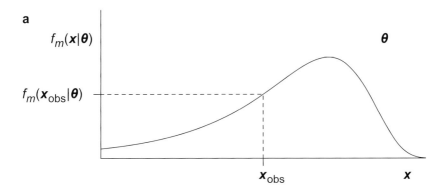

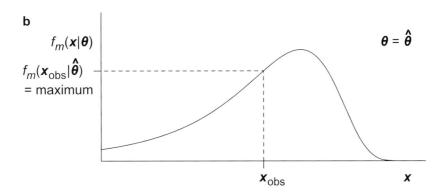

Figure 3.3 Maximum likelihood principle. (a) General situation (parameter vector θ). (b) The maximum likelihood estimate, $\hat{\theta}$, maximizes the PDF, f_m, of the m-dimensional data vector, x, at the observation point, x_{obs}. In this treatment, the data are considered as fixed and the parameter as variable. The function $L(\theta) = f_m(x_{\text{obs}}|\theta)$ is called likelihood function. See also Box 3.1.

likelihood is to take that θ-value as estimator, for which the observed data would have had the highest probability of being generated by the system.

Amazingly, maximum likelihood's treatment of the data as fixed and the parameter as variable, is mirrored in another estimation technique. The Bayesian approach uses the data to update the knowledge about a parameter, to move from the prior distribution of θ to the posterior distribution.

I personally have to confess that the least-squares estimation technique has always appeared intuitively clear and appealing to me: take as estimate that θ-value, for which the deviation between data and fit is smallest. This can be done, for example, in linear regression (Section D.1.2). My eyebrows raise when I notice attempts by one or the other camp of professional statisticians to convince us that for "philosophical reasons" one estimation approach is superior to another. They do so usually without detailing what kind of philosophy that is. Scientific crusades. Boring.

Rather, let us follow a pragmatic approach to estimation. As with all applied scientists, we climatologists have the freedom to select which estimator to use, even for such simple tasks as estimating the standard deviation. As noted in that Reflection (Box D.1), we can run a simulation experiment, or resort to existing Monte Carlo evidence about which procedure achieves the best estimation. Informed by this evidence, we apply the optimal technique to our precious data.

Ronald Aylmer Fisher (1890–1962) was the man who single-handedly worked out maximum likelihood estimation in the 1920s (Fisher 1922, 1925). As so often in the history of science, the concept had been mentioned by others before (in the century before Fisher). It is impossible for a researcher who works with data to miss Fisher's name since he left his mark in many diverse branches of statistical analysis. I myself met him first in my PhD studies, when I looked for tests for the number of peaks in spectral analysis (Fisher 1929). For writing the chapter on correlation (where one climate variable, X, is related to another, Y) of my book on climate time series analysis, I referred to his derivation of the distribution of the correlation coefficient (Fisher 1915) and the z-transformation (Fisher 1921). As regards the context of the present textbook on extremes, he contributed a benchmark paper (Fisher and Tippett 1928) to the emergence of the GEV distribution for the block maximum (Figure 2.7). It seems to me that the foundation stone of his work was maximum likelihood, on which he built in the various estimation problems he encountered in his applied work (e.g., at Rothamsted Experimental Station in the United Kingdom, where agricultural research had been performed since 1843). His hard-working life and his scientific rivalries, which seem to root in his search for truth, are covered in his biography by his daughter (Box 1978).

3.1.4 Heavy-Tail Distributions

The tail probability for a GP distribution (Eq. 3.5) is in the general case ($\xi \neq 0$) given by

$$p = 1 - F_{GP}(x_{ext}) = \left\{ 1 + \xi\,(x_{ext} - u)\,\big/\left[\sigma + \xi\,(u - \mu)\right] \right\}^{-1/\xi}. \qquad (3.9)$$

This formula means that for large x_{ext}-values, p is approximately given by the proportionality

$$p \propto x_{ext}^{-1/\xi}. \qquad (3.10)$$

This is a power law, and the type of tail is called Pareto tail. The power law means that p does not decrease very fast (e.g., exponentially) with x_{ext}, and there is a considerably high probability for observing large extremes. The Pareto tail is, hence, called heavy.

The GP distribution applies to POT extremes. The GEV distribution applies to block extremes. Also the GEV distribution (Eq. 3.1) shows approximately a Pareto tail, described by the parameter ξ.

Risk analysis deals with the tail behaviour. For this purpose, a robust estimation approach may be based on the fit of a heavy-tail distribution only to the extremal part of a distribution of data,

$$p \propto x_{ext}^{-\alpha}. \qquad (3.11)$$

This formula is assumed to hold above a threshold value, $x_{ext} > u \geq 0$. We have replaced $1/\xi$ from Eq. (3.10) by α, which is conventionally called heavy-tail index parameter. The robustness stems from the fact that the heavy-tail distribution model may apply to a fairly wider class of extreme values, not just the GEV or GP types. It may, for example, be still useful in cases where the GEV approximation and the regularity conditions (Section 3.1.1) are not exactly fulfilled and a GEV fit could lead to biased results. This means that in risk analysis a heavy-tail distribution may constitute a powerful estimation alternative to the GEV or GP distributions.

A classic estimator of the heavy-tail index parameter on the basis of a set of data was devised by Hill (1975). Let us consider the right tail (the largest values). This assumption does not loose generality because the left tail can be analyzed in the same manner after the data values have been multiplied by -1. A method that is general is appealing since it can be applied in many situations in a consistent manner. Let us for notational convenience omit to write the subscript "ext" and denote the dataset as $\{x(i)\}_{i-1}^{n}$. Let further x' denote the x-values sorted according to descending size, $x'(1) \geq x'(2) \geq \cdots \geq x'(n)$. Let us assume zero mean, which can be achieved by means of subtraction of the sample mean, $\bar{x} = \sum_{i=1}^{n} x(i)/n$, from the data. Finally, let there be $K \geq 2$ positive x'-values. The Hill estimator is then

$$\widehat{\alpha}_k = k \left[\sum_{i=1}^{k} \ln \frac{x'(i)}{x'(k+1)} \right]^{-1}. \qquad (3.12)$$

The parameter $k \leq K - 1$ is denoted as order parameter. If $K < 2$, then the Hill estimator cannot be applied. The selection of k completes the estimation ($\widehat{\alpha}$ is given by $\widehat{\alpha}_k$ for the selected k). Order selection constitutes a statistical trade-off problem (Hill 1975). Large k leads to usage of many data points and a small estimation variance. However, the danger is then that points are included for which Eq. (3.11) does not hold (i.e., bias). On the other hand, small k leads to a small estimation bias and a large variance. Order selection has been, hence, called the "Achilles' heel" of heavy-tail index parameter estimation (Resnick 2007).

Mudelsee and Bermejo (2017) introduced an order selector that is optimal in the sense that it minimizes an error measure that is thought to be representative of the trade-off problem. The minimization is data-adaptive and carried out by means of a brute-force search (Appendix E.1). It is therefore a computationally intensive approach.

An artificial time series (Figure 3.4a), drawn from a heavy-tail distribution with prescribed index parameter, $\alpha = 1.5$, serves to illustrate the approach. More precisely, the prescribed distribution is called stable distribution (Nolan 2003), for which the index parameter describes both tails, although only the right tail is analyzed. No persistence is prescribed, although the approach (Mudelsee and Bermejo 2017) can deal also with that.

The sequence $\widehat{\alpha}_k$, when plotted against the order parameter, k, indicates a weak downward trend (Figure 3.4b). This means that the order-selection recipe to look for which k-values the sequence has a plateau (Resnick 2007) is not helpful.

However, the error measure for $\widehat{\alpha}_k$ shows a clearly expressed minimum at $k_{\text{opt}} = 1024$ (Figure 3.4c). This means that the optimal order selector advises to utilize 1024/5000 or about 20.5% of the largest values for the estimation.

The resulting estimate is $\widehat{\alpha} = 1.55$. The fitted heavy-tail distribution agrees for this estimate well with the histogram (Figure 3.4d). An error bar of type

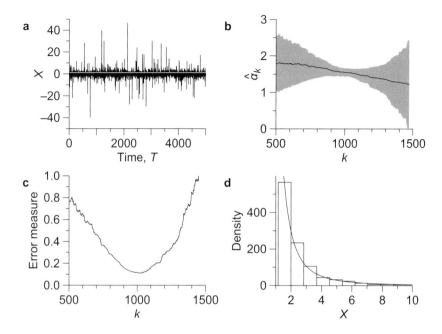

Figure 3.4 Optimal heavy-tail index parameter estimation. (a) Time series, prescribed parameters: $n = 5000$ and $\alpha = 1.5$. Here the heavy-tail index parameter describes both tails, although only the right tail is analyzed. Also shown (gray line) is the value $x'(k_{opt}) = 1.2$ (see c for k_{opt}) for the right tail. (b) Sequence $\widehat{\alpha}_k$ (Eq. 3.12) in dependence on order parameter, k. Also shown (shaded) is the error measure for $\widehat{\alpha}_k$ around the sequence. (c) Error measure for $\widehat{\alpha}_k$. The curve has a minimum at $k_{opt} = 1024$, which leads to an estimate of $\widehat{\alpha} = 1.55$. (d) Density estimates in form of a histogram and a power law for the tail probability (solid line). The latter curve is scaled such that the area under the curve ($x \geq x'(k_{opt})$) agrees with the number of extreme events. The upper bound for the X-axis is set at a value of 10 for better legibility.

RMSE (Section D.1) is determined by means of 100 parametric Monte Carlo simulations (Mudelsee and Bermejo 2017). The simulations consist in random draws of time series from a stable distribution with prescribed α equal to $\widehat{\alpha}$, to which each time the optimal order selector is applied to produce simulation copies of $\widehat{\alpha}$. The resulting error bar for the estimate of $\widehat{\alpha} = 1.55$ equals 0.06, which means that the estimate is not significantly different from the prescribed value, $\alpha = 1.5$.

See Exercise 3.3 for replicating this heavy-tail index parameter estimation and analysing further time series.

3.2 Nonstationary Poisson Process

Nonstationarity means that the properties of the data generating process vary over time. This is relevant in the context of this book because climate change may lead to changes in the extremal part of a distribution of a climate variable.

One seemingly straightforward option is the extension of the stationary models by means of a time dependence in the parameters. For example, the GEV model (Section 3.1.1) can be reformulated as

$$\mu(T) = \beta_0 + \beta_1 T, \tag{3.13}$$

$$\sigma(T) = \exp(\gamma_0 + \gamma_1 T), \tag{3.14}$$

$$\xi(T) = \delta_0 + \delta_1 T. \tag{3.15}$$

(The exponential function ensures a positive scale parameter.)

However, there are two problems with the extension of the parametric model. First, there is a technical challenge. The maximization of the log-likelihood function (Section E.1) becomes numerically difficult. This is owing to the log-likelihood function, which now depends on six parameters. Second, there is a principal limitation. There is no guarantee that the true time dependence does follow the simple linear forms of Eqs. (3.13)–(3.15). This limitation may be more severe for longer records, where the time dependence can be more complex.

A more flexible option to describe the nonstationarity in the extremal part is the usage of a Poisson process. This is a statistical model for the occurrence of events. The stationary or homogeneous Poisson process has one parameter: the occurrence rate, λ, is defined as the number of independent events per time unit. We extend this by writing a time dependence, $\lambda(T)$, and call this model a nonstationary Poisson process. In the literature, it is also denoted as inhomogeneous Poisson process (Cox and Lewis 1966).

3.2.1 Kernel Occurrence Rate Estimation

Let us assume our data are in the form of event times (Section 2.1). How can the data, $\{T_{\text{ext}}(j)\}_{j=1}^{m}$, be utilized to estimate the occurrence rate, $\lambda(T)$?

The mathematical formula for the kernel estimator is

$$\widehat{\lambda}(T) = h^{-1} \sum_{j=1}^{m} K\left(\left[T - T_{\text{ext}}(j)\right]/h\right), \tag{3.16}$$

where h is the bandwidth and K is a kernel function. We take the Gaussian kernel, $K(y) = (2\pi)^{-1/2} \exp(-y^2/2)$.

The idea behind the kernel approach is to count the number of events in a continuously shifted time window (Figure 3.5). This yields a full curve as estimated occurrence rate, not just a few points as in case of contiguous windows. Using smooth (e.g., Gaussian) kernels results in smooth occurrence rates, which have better mathematical properties (bias, standard error) – and also a nicer look – than unsmooth estimates based on uniform kernels.

3.2.2 Bandwidth Selection

More important than the choice of the kernel type is the selection of the bandwidth, h. A large bandwidth means that many data points effectively contribute to the estimation. This has the advantage of small standard errors (Section D.1). However, a large bandwidth means also that fine details may be obscured, corresponding to an increase of estimation bias. On the other hand, a small bandwidth leads to a reduced bias and larger standard errors. This is the smoothing problem, which we have already met in another context (Exercise 2.3). The effects of bandwidth selection are illustrated in Figure 3.6.

As a solution of the smoothing problem in our context of kernel occurrence rate estimation, Brooks and Marron (1991) developed the cross-validation bandwidth selector. This is the minimizer of the function

$$C(h) = \int \left[\widehat{\lambda}(T)\right]^2 dT - 2 \sum_{j=1}^{m} \widehat{\lambda}_j\big(T_{\text{ext}}(j)\big), \qquad (3.17)$$

where

$$\widehat{\lambda}_j(T) = \sum_{k=1,\ k \neq j}^{m} h^{-1} K\big(\left[T - T_{\text{ext}}(k)\right]/h\big) \qquad (3.18)$$

is the delete-one estimate.

The integral (Eq. 3.17) is calculated over the observation interval. If the events have been detected on an observed time series (Section 2.4), then the observation interval is given by $[t(1); t(n)]$. If the data are of type "event times," then additional knowledge has to be consulted.

In the case of the heavy Elbe winter floods (Figure 2.1b), the observation interval is [1020; 2002.333]. The lower bound is due to where the constructors of the flood record (Mudelsee et al. 2003) started the analysis of the source texts. The upper bound, which corresponds to the date 30 April 2002, reflects the information available at completion of that work. The cross-validated bandwidth of 41 a (Figure 3.7) is used for occurrence rate estimation (Figure 3.6c).

Methods

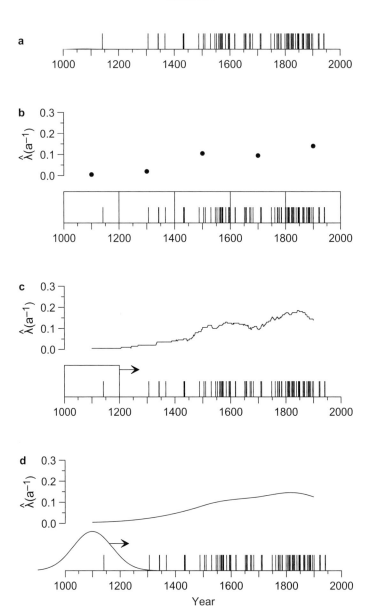

Figure 3.5 Kernel occurrence rate estimation. (a) Event times for the heavy
Elbe winter floods (Figure 2.1b). (b) Occurrence rate estimate (upper part, filled
symbols) based on the number of events in contiguous 200-year windows (lower
part). (c) Occurrence rate estimate (upper part, solid line) based on the number of
events in continuously shifted 200-year windows (lower part). This window type
is called a uniform kernel. (d) Occurrence rate estimate (upper part, solid line)
based on a Gaussian kernel with bandwidth 100 years (lower part).

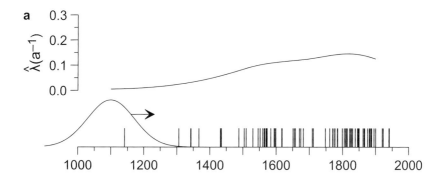

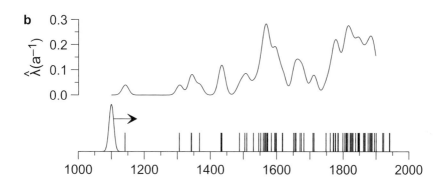

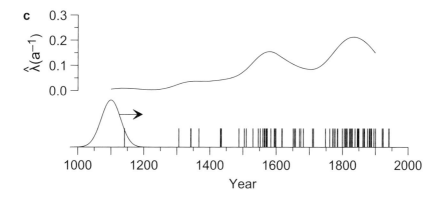

Figure 3.6 Bandwidth selection. Occurrence rate estimate for the heavy Elbe winter floods (Figure 2.1b) based on a Gaussian kernel with bandwidth 100 years (a), 10 years (b), and 41 years (c).

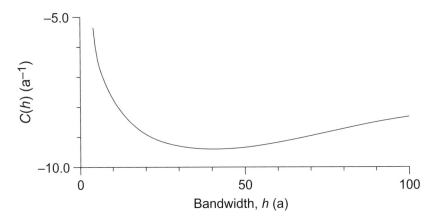

Figure 3.7 Cross-validation function. The function $C(h)$ (Eq. 3.17) for the heavy Elbe winter floods (Figure 2.1b) has a minimum at $h \approx 41$ a.

3.2.3 Boundary Bias Correction

The extreme event times have been found within an observation interval, for example, $[t(1); t(n)]$. We are certainly also interested in the occurrence rate, $\lambda(T)$, at the interval boundaries. However, at the boundaries half of the kernel window cannot collect events because there are no events observed outside (Figure 3.5). The result would be an underestimation of $\lambda(T)$ near the boundaries, that is, a negative bias.

An elegant way to correct for the boundary bias is to generate artificial pseudodata outside of the observation interval. It can be shown that this corresponds to a simple extrapolation of the distribution of the event times (Cowling and Hall 1996). The kernel estimation of the occurrence rate (Eq. 3.16) and the cross-validation function for bandwidth selection (Eqs. 3.17 and 3.18) are then calculated on the event times augmented by the pseudodata. The simplest form of pseudodata generation is reflection at the interval boundaries, shown for the right pseudodata (upper bound) in Figure 3.8.

How many pseudodata to generate? Figure 3.8 shows that an extension into the outside by three bandwidths is sufficient because then the kernel contribution is already small.

3.2.4 Confidence Band Construction

A measure of the uncertainty of the occurrence rate estimation is indispensable for interpreting the result. It helps for assessing, for example, whether the

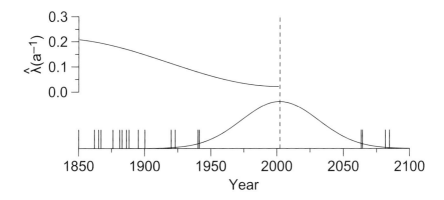

Figure 3.8 Pseudodata generation. The right pseudodata (lower part, on the right) for the heavy Elbe winter floods (Figure 2.1b) are generated by reflecting the event times (lower part, on the left) at the upper bound of the observation interval (dashed line) to the right. The occurrence rate (upper part) is estimated by means of the kernel and utilization of the event times augmented by the pseudodata.

ups and dows in the curve are statistically significant. Figure 3.9 shows the construction of a confidence band.

Key to the construction are computer-simulated copies of the data generating Poisson process. The copies are called bootstrap resamples and generated by randomly selecting, with replacement, event by event from the original set, $\{T_{\text{ext}}(j)\}_{j=1}^{m}$, augmented by the pseudodata. The set of a resample has the same data size as the original data plus pseudodata. The random selection means that it is possible to draw some events several times and miss other events.

The kernel occurrence rate estimation is repeated on a resample, which results in a simulated estimation curve (Figure 3.9b). The procedure of resampling and re-estimation is repeated many times, until typically 2000 simulated curves are available.

The simulated occurrence rate curves are used to construct the confidence band. The band is obtained by a concatenation of the confidence intervals (Section D.1) over time. A percentile interval would result from the percentiles of the distribution of the simulated occurrence rates. For example, from 2000 simulations the 1900th largest and 100th largest simulated occurrence rate values define the 90% percentile interval. Instead of the percentile, we employ the so-called percentile-t confidence band, which involves further processing of the simulations (Mudelsee 2014: algorithm 6.1 therein), because this has been shown to be more accurate (Cowling et al. 1996). The resulting confidence band is pointwise, that is, for each time value, T, at which the

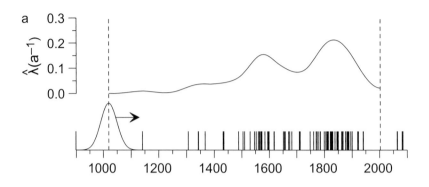

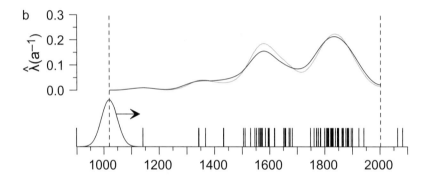

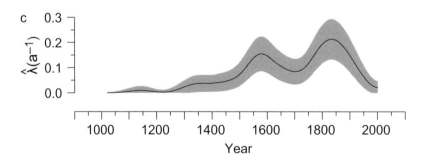

Figure 3.9 Confidence band construction. (a) Occurrence rate estimate (upper part, solid black line) based on a Gaussian kernel with bandwidth 41 years and the event times (Figure 2.1b) augmented by right (Figure 3.8) and left pseudodata (lower part). (b) Occurrence rate estimate (upper part, solid black line) and simulated estimate (upper part, solid gray line) based on a bootstrap resample (lower part) of the data. (c) Occurrence rate estimate (solid black line) and 90% confidence band (shaded) that is constructed on the basis of 2000 bootstrap resamples.

occurrence rate is estimated (Eq. 3.16), the probability that the band includes the true occurrence rate value is nominally 90%.

3.2.5 Cox–Lewis Test

It is possible to formulate a parametric model for the occurrence rate. Since $\lambda(T)$ cannot be negative, it is convenient to employ the exponential function. A particularly simple model is

$$\lambda(T) = \exp{(\beta_0 + \beta_1 T)}. \tag{3.19}$$

This is a monotonic function. The model serves as a simple description of an increasing or decreasing trend of the occurrence rate. It does not offer the flexibility of the nonparametric kernel approach. However, it is suited to a situation where the task is not quantification of $\lambda(T)$ but rather testing whether $\lambda(T)$ shows an increase or decrease.

To enter the statistical area of hypothesis testing (Section D.2), we consider the null hypothesis H_0: "constant occurrence rate ($\beta_1 = 0$)" and the alternative H_1: "increasing occurrence rate ($\beta_1 > 0$)." Cox and Lewis (1966) devised a test statistic, U_{CL}, which guides us in our decision whether or not to reject H_0 against H_1:

$$U_{\text{CL}} = \frac{\sum_{j=1}^{m} T_{\text{ext}}(j)/m - [T(n) + T(1)]/2}{[T(n) - T(1)](12m)^{-1/2}}. \tag{3.20}$$

Under H_0, the distribution of U_{CL} approaches rapidly with increasing m a standard normal shape. This facilitates the calculation of the P-value (Section D.2). The idea of the test is to compare the average of the event times with the center of the observation interval, $[T(1); T(n)]$. On the sample level, plug in the values $\{t_{\text{ext}}(j)\}_{j=1}^{m}$, $t(1)$, and $t(n)$ to obtain the test statistic, u_{CL}. If the data are of type "event times," then additional knowledge about the collection of the data is used to determine the observation interval. The other alternative H_1: "decreasing occurrence rate ($\beta_1 < 0$)" can be tested analogously.

Box 3.2 **Personal Reflection: Efron, Hall, and the bootstrap**

Do you remember the confidence band construction in Section 3.2.4? We had to simulate from the data generating process but we did not know the distributional shape. The key was to "let the data speak for themselves" and randomly select, with replacement, event by event from the original dataset, until we had drawn a resample of same data size. Then we re-estimated the occurrence rate using the resample. This yielded a simulated estimation curve. This procedure was repeated many times. This yielded many different simulated estimation curves. Finally, we constructed the confidence band from the curves.

It may appear as magic that this resampling method produces reliable results. However, Monte Carlo tests have shown that it works. Attributed to Karl Friedrich Hieronymus Baron von Münchhausen (1720–1797) are many magic stories, which are quite famous in Germany. In one of those, he claimed to have rescued himself from sinking into a swamp by pulling himself up by his own hair. American or English readers may be reminded of the saying "to pull oneself over a fence by one's bootstraps." Anyhow, our method is called bootstrap resampling.

Bradley Efron (born 1938) is the man behind the bootstrap. In his famous paper (Efron 1979), he collected and synthesized earlier attempts by others and showed theoretically on simple estimation problems the correctness of the bootstrap. He and Peter Hall (1951–2016) generated in the subsequent years much theoretical and computer experimental knowledge about bootstrap uncertainty measures, which found entrance into the canon of statistical estimation (Section D.1).

Science is being performed by humans, and humans are not perfect rationalists. The road to success the bootstrap was forced to take in the land of statistics: it was not easy. Today, it is one major component in computationally intensive, data-driven statistical methodology (Efron and Hastie 2016).

I myself heard about the bootstrap when working on my PhD dissertation, but I learned about its adaptation to time series (i.e., persistence) only at the end of the 1990s. This was when I was a postdoc at the University of Kent, where I had the luck to meet another postdoc, Cees Diks, who showed me one paper (Politis and Romano 1994). With the two mentioned giants in statistics, I only had mini exchanges of e-mails. While I have not yet met Efron, I once followed an oral presentation by Hall. Learning about his death and discovering the memorial page set up by Steve Marron (http://marron.web.unc.edu/sample-page/peter-hall-memorial-page/, 9 December 2017) left a lasting impact on me. Sometimes I feel ashamed when I apprehend the greatness of contemporary researchers and colleagues only after their death.

Exercises

Exercise 3.1: Return Level

The return level (Section 3.1.3) for a GEV distribution with $\xi \neq 0$ is given by

$$x_p = \mu - (\sigma/\xi) \left\{ 1 - \left[-\ln(1-p) \right]^{-\xi} \right\}. \tag{3.21}$$

The standard error of the estimated return level is equal to the square root of its variance,

$$se_{\widehat{x}_p} = [VAR(\widehat{x}_p)]^{1/2}. \tag{3.22}$$

The variance can, in turn, be approximately calculated from the variances and covariances of the GEV parameters μ, σ, and ξ as

$$VAR(x_p) \approx \left(\frac{\partial x_p}{\partial \mu} \right)^2 VAR[\mu] + \left(\frac{\partial x_p}{\partial \sigma} \right)^2 VAR[\sigma] + \left(\frac{\partial x_p}{\partial \xi} \right)^2 VAR[\xi]$$

$$+ 2 \left(\frac{\partial x_p}{\partial \mu} \right) \left(\frac{\partial x_p}{\partial \sigma} \right) COV[\mu, \sigma] + 2 \left(\frac{\partial x_p}{\partial \mu} \right) \left(\frac{\partial x_p}{\partial \xi} \right) COV[\mu, \xi]$$

$$+ 2 \left(\frac{\partial x_p}{\partial \sigma} \right) \left(\frac{\partial x_p}{\partial \xi} \right) COV[\sigma, \xi]. \tag{3.23}$$

This approximation stems from a series expansion of the variance in terms of powers of small deviations. Such an error expansion type is called the delta method or error propagation.

The covariance operator, COV, is defined on two random variables, X and Y, as

$$COV[X, Y] = E\big[(X - E[X]) \cdot (Y - E[Y])\big], \tag{3.24}$$

where E is the expectation operator (Exercise 2.4). A special case of this formula is $COV[X, X] = VAR[X]$.

The exercise consists in the calculation of the variance of the estimated return level (Eq. 3.23) by inserting the partial derivatives. For example, $(\partial x_p / \partial \mu)$ is the derivative of x_p with respect to μ. The derivatives have to be calculated from Eq. (3.21). In practice, on the sample level, the "hats" have to be inserted in Eq. (3.23). Furthermore, the parameter estimates and the estimated values for the variance and covariance terms have to be plugged in on the right-hand side of Eq. (3.23). These estimates of parameters, variances, and covariances are an output product of the maximum likelihood technique (Exercise 3.2).

Exercise 3.2: Elbe Summer Floods

The first task is to replicate the GEV estimation for the Elbe summer floods (Figure 3.2).

The monthly maxima data (i.e., block length one month) are given, together with monthly averages and minima, in the file Elbe-S-M1.txt (Section F.1). Use this file to construct an input data file, which consists of $\{t_{\text{ext}}(j), x_{\text{ext}}(j)\}_{j=1}^{m}$. See Section 2.3 for the input data format; note that for notational convenience we omit to write a prime symbol.

Use the monthly maxima data to generate further input data files (block length two, three, and six months).

For estimation of the GEV parameters and the return level, use the software GEVMLEST (Section F.2).

For the plot of the estimated GEV density function (Figure 3.2c), take the first derivative of the distribution function (Eq. 3.1) and plug in the parameter estimates. For the histogram of the block maxima (Figure 3.2c), divide the interval from the smallest to the largest x_{ext} value into a number of

$$NINT\left[\left[\max\left(x_{\text{ext}}(j)\right) - \min\left(x_{\text{ext}}(j)\right)\right] m^{-1/3} / (3.49\, s_{n-1})\right], \qquad (3.25)$$

classes of equal width. In this formula, $NINT$ is the nearest integer function and s_{n-1} is the sample standard deviation with denominator $n-1$ (Eq. D.19) on the sample level. Although this class-width selector (Scott 1979) was designed for Gaussian distributions, it works well enough here for the comparison of the histogram with the GEV density.

The second task is to play with the estimation parameters (tail probability and block length) to acquire an intuition for the estimation. Observe how the error for the return level (Exercise 3.1) increases with the prescribed return period (Figure 3.2d). What is the return level with standard error for a 10,000-year event based on two-monthly block maxima? Can this value, which is based on barely more than 200 years of data, be meaningfully interpreted?

Exercise 3.3: Heavy-Tail Index Parameter Estimation

This is a software exercise using the program HT. See Section F.2 for access to the software and its manual. Consult the manual ("Getting Started") for the installation of HT on your computer.

The first part of the exercise is to replicate the analysis shown in Figure 3.4. Read the manual (Section 1 therein) to become able to work interactively with HT (i.e., data generation and parameter estimation).

Generate a time series with the same properties ($n = 5000$, stable distribution with $\alpha = 1.5$) as in Figure 3.4a. The persistence time to be prescribed for the data generation is zero. That means, not Eq. (2.8) is used (since you cannot divide by zero), but a purely random process with stable distributed innovations.

Estimate α using the Hill estimator (Eq. 3.12) and a brute-force order selector. Determine an error bar for $\widehat{\alpha}_k$ using Monte Carlo simulations. This is a computationally intensive step, and HT therefore employs parallel computing (Section E.2) for it. The software further has a quasi-brute-force order selector, described in the manual (Section 1 therein), which may speed up the estimation.

The second part of the exercise is to play with the parameter settings for data generation and estimation to acquire an intuition for working with HT.

The third part is to perform a reanalysis of heavy-tail index parameter estimation on the daily runoff time series from the river Elbe at station Dresden for the hydrological summer for the interval from May 1806 to October 2013 (Figure 3.2a). The time series is given in the file Elbe-S.txt (Section F.1). Which value for the optimal order, k_{opt}, and which estimates of τ and α (with error bars) do you get? Compare your results afterwards with those obtained by Mudelsee and Bermejo (2017: section 6 therein).

Exercise 3.4: Heavy Elbe Winter Floods and the Cox–Lewis Test

The heavy Elbe winter floods show a downward trend in the occurrence rate since the beginning of the nineteenth century, which is assessed as highly significant by means of the confidence band (Figure 3.9c). Use the Cox–Lewis test (Section 3.2.5) to test the significance on basis of the data from the instrumental period, which was, at the time of performing the calculations for the paper (Mudelsee et al. 2003), the interval from 1 January 1850 to 30 April 2002 (the end of the hydrological winter). That means, calculate the test statistic, u_{CL}, and look up (in a textbook or on the internet) the one-sided P-value. The data are included in the file Elbe-W23.txt (Section F.1).

Reading Material

Coles (2001) is a textbook on extreme value analysis with a focus on stationary processes. Cox and Lewis (1966) is a textbook on Poisson processes. Efron and Hastie (2016) is a book on computer age statistical inference. Resnick (2007) is an accessible textbook on heavy-tail distributions.

Efron (1979), a classic paper, introduces bootstrap resampling. Hill (1975) introduces a classic estimator of the heavy-tail index parameter. Mudelsee et al. (2003) introduce bootstrap resampling to flood risk analysis. Scott (1979) presents a class-width selector for histograms.

4

Floods and Droughts

We apply the methods from Chapter 3 in two case studies to data presented in Chapters 1 and 2. While the floods of the river Elbe during the past millennium have been be studied in several ways (Section 4.1), the monsoon droughts during the Holocene are less well understood (Section 4.2). Two case studies certainly cannot exhaust all aspects of floods and droughts. Therefore, we indicate current research directions (Section 4.3). Finally, we give a short summary for risk analysts and other people who focus on the current state and the near future of floods and droughts (Section 4.4).

4.1 Case Study: Elbe Floods

The Elbe is a major river in central Europe, which drains an area of ~148,000 km². An exceptionally strong flood occurred in August 2002, which cost overall losses of 11.7 billion EUR (not inflation-adjusted) in Germany alone (Ernst Rauch, Munich Re, Germany, personal communication, 2014). Other exceptionally strong floods occurred before and after that event, for example, in June 2013 (Conradt et al. 2013) or March–April 1845 (Königliche Elbstrombauverwaltung 1898). Knowledge about Elbe flood risk has thus a strong relevance for society and the economy.

Previous chapters presented the Elbe flood data (Chapter 2) and some statistical results (Chapter 3). Data and results are, strictly speaking, representative only for the middle Elbe (Figure 4.1). Here we revisit data and results, but focus on the interpretation of the statistical findings. We carry out sensitivity analyses to evaluate climatological and other forcing factors of trends in flood occurrence rate. The overall goal is to assess the reliability and robustness of results from a climatological–physical perspective.

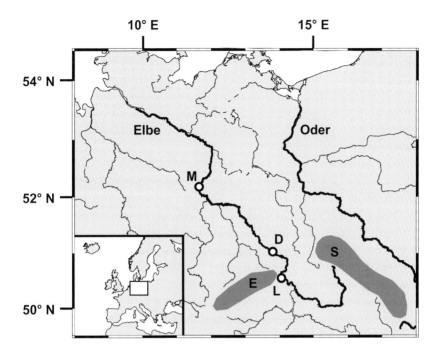

Figure 4.1 Map of the region of the river Elbe. The region is indicated by the white box in the inset. The neighboring river Oder is also shown. The middle Elbe is between Litoměřice (L) in the Czech Republic and Magdeburg (M) in Germany. Elbe runoff measurements (Figure 3.2) are from the station in Dresden (D) in Germany. Elbe and Oder hydrology is influenced by low-range mountainous climate in the Erzgebirge (E) and Sudeten (S) mountains.

4.1.1 Data Quality

The Elbe flood data, shown for the winter in Figure 2.1, cover the instrumental period (back to approximately AD 1850), for which runoff values are available, and the earlier period (back to approximately AD 1000), for which documentary data exist. The flood events for the instrumental period were obtained by placement of a magnitude threshold (Figure 2.2) to independent flood peaks in runoff.

Runoff is usually inferred via water stage and a stage–runoff calibration (Figure 2.2). Since a flood event can alter the geometry of a river, it can also alter the numerical values of the calibration curve. The data quality for the instrumental period depends therefore on (1) the accuracy of the calibration, (2) how stable the stage–runoff relation is over time, and (3) how frequently the calibration has been updated. The Elbe runoff record from Dresden, on which the flood events (Figure 2.1) are mainly based, fulfills these data-quality requirements to an excellent degree (Mudelsee et al. 2004). Since the stage–runoff calibration shows a slow increase over time and updating was less frequent before 1960, the earlier part of the flood record may be influenced by bias. However, the magnitude classes (Figure 2.2) are much wider than any statistical or systematic uncertainty associated with the flood runoff values. The Elbe flood record for the instrumental period is therefore of excellent data quality.

The instrumental runoff record used by Mudelsee et al. (2003) for the reconstruction of Elbe floods for winter (Figure 2.1) and summer has been later extended back in time from 1850 to 1806 by means of data supplied by the Global Runoff Data Centre (Koblenz, Germany, personal communication, 2013–2014). The application of the magnitude classification curve (Figure 2.2) to the new runoff data yields a new flood record for the interval from 1806 to 1850. A comparison with the existing flood record (Mudelsee et al. 2003) allows us to assess the data quality in a quantitative manner (Table 4.1).

It turns out that the two exceptionally strong events (magnitude class 3) are recorded both by documentary and runoff data, while there are clearly more class-2 events inferred from the documents (Table 4.1). The occurrence rates for heavy events (class 2–3) inferred from the new runoff data should therefore be smaller than the rates (Mudelsee et al. 2003) from the documentary data for the interval [1806; 1850]. This is also true when the interval is extended to both sides by a few decades because of the kernel bandwidth of 41 a used for the estimation (Figure 3.6c). However, the rates are not significantly smaller, outside of the confidence bands as the analysis shows (Figure 4.2a, b). The minor events (class 1) reveal a good agreement, with some events recorded

Table 4.1. *Comparison of the number of Elbe flood events for the interval from 1806 to 1850. The entries show in dependence of the magnitude class the events recorded in the documentary sources (Mudelsee et al. 2003) and/or in the runoff values.*

Class	Documentary and runoff	Documentary, not runoff	Runoff not documentary
3	2	0	0
2	7	5	0
2–3	9	5	0
1	20	12	8

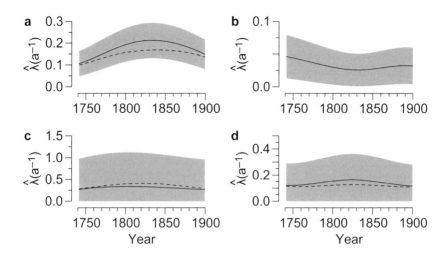

Figure 4.2 Comparison of the occurrence rates of Elbe floods for the interval from 1806 to 1850. Each panel (a, winter floods, magnitude class 2–3; b, summer floods, magnitude class 2–3; c, winter floods, magnitude class 1; d, summer floods, magnitude class 1) shows $\widehat{\lambda}(T)$ (solid line) with 90% confidence band (shaded) that is constructed with a bandwidth of 41 a on the basis of 2000 bootstrap resamples from the documentary data for the interval [1806; 1850]. Also shown are the occurrence rates (dashed lines) estimated from the runoff data for that interval. (In panel b, documentary- and runoff-derived occurrence rates agree with each other, and the curves overlap.)

just by one or the other data type (Table 4.1). Since the absolute numbers (20 + 12 = 32 and 20 + 8 = 28) are not so different, the effects on the estimates of the occurrence rate are small (Figure 4.2c, d). Table 4.1 also shows that document loss is not a problem for the interval [1806; 1850]. It is not clear whether the documentary- or the runoff-derived flood record is closer to the truth.

The data quality for the interval before the instrumental period back to AD 1500 has been assessed as generally good (Mudelsee et al. 2004). This is owing to the availability of considerably more sources about an event than in earlier time periods and, in particular, the existence of a stone bridge (with water-stage marks) in Dresden. The risk of document loss is small, especially for heavy floods. There has been criticism of the quality of the documentary data, see the personal reflection about the Weikinn source texts (Box 2.1). Therefore, Mudelsee et al. (2003) produced additional flood records using the compilation by the historian Militzer, which covers the interval from 1500 to 1799. It turned out that for class 2–3 events, the re-estimated flood occurrence rates in the overlapping time interval were indistinguishable (within the confidence band) from Weikinn's rates, whereas for class 1–3 events, Militzer's rates were lower (Mudelsee et al. 2004: figure 8 therein). Therefore, the interpretation for minor flood events (class 1) requires caution.

The data quality for the period before ~1500 is clearly reduced. Mudelsee et al. (2003, 2004) concluded that document loss hampers the reconstructed Elbe flood record, especially for minor events (class 1). The reasons are given in Box 2.1: few printed books existed and people had other occupations than to record by hand a flood event. In such a situation, natural climate archives (Appendix B) can help to generate a record that is less influenced by human bias (Wilhelm et al. 2019).

4.1.2 Reservoirs

Reservoirs along a river are built for supply of drinking water and management of water stage. The combined size of all reservoirs above the Elbe station Dresden has increased since the beginning of the twentieth century to a present level of 237×10^6 m^6 (Figure 4.3). This value can be utilized for the management of winter floods (via release of water prior to a flood). The manageable size for the summer is slightly smaller.

Is the observed, statistically significant downward trend in occurrence of heavy Elbe winter floods since the early nineteenth century (Figure 3.9c) robust against reservoir-size influences? Or is it an artifact due to the increasing ability (Figure 4.3) to manage runoff and cut off a flood peak?

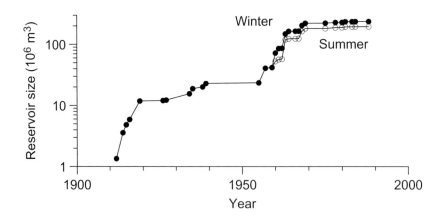

Figure 4.3 Reservoir size above Elbe station Dresden. Data from Bundesanstalt für Gewässerkunde (2000)

Table 4.2. *Cox–Lewis test results on reservoir-size-corrected Elbe floods for the interval from 1852 to 2002. The test (Section 3.2.5) of the null hypothesis "constant occurrence rate" (no trend) is performed at a one-sided significance level of 90%.*

Season	Magnitude class	Trend	
		Uncorrected	Corrected
Winter	2–3	Down	Down
	1–3	Down	No
Summer	2–3	No	No
	1–3	No	Up

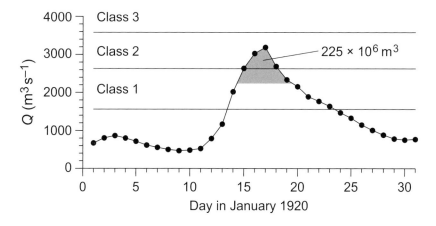

Figure 4.4 Reservoir-size correction of a flood peak. The January 1920 winter flood peak of the Elbe at Dresden could have been reduced from a class-2 to a class-1 event if at that time the present manageable reservoir size, which is 225×10^6 m^6 larger, were available.

To answer these questions, Mudelsee et al. (2003) constructed flood records that would have resulted if reservoir size were constant at the present level. A corrected, reduced flood peak results from "cutting off" the difference in reservoir size (Figure 4.4). This exercise had to be carried out for the full instrumental period.

The results, explained by Mudelsee et al. (2004: p. 15 therein), show that in the case of class 2–3 floods, no changes in the sign and the confidence level of the trend could be induced by flood management of reservoirs (Table 4.2). The reservoir size is too small for influencing the occurrence of heavy floods.

When including also class-1 floods, a significant reduction of flood risk is theoretically possible. However, such a reduction assumes 100% utilization of the available reservoirs – which is practically impossible. This means that the theoretical reduction in flood size is an upper limit. The consequence is that the trend test applied to reservoir-size corrected records becomes liberal (that is, it has a lower than nominal confidence level) as regards upward trends, and it becomes conservative (higher than nominal confidence level) as regards downward trends. Therefore, the upward trend in reservoir-size corrected Elbe summer floods for class 1–3 (Table 4.2) is likely an artifact owing to an over-correction.

4.1.3 Further River Training

Building of reservoirs (Section 4.1.2) is one type of river training. Further types of work on the middle Elbe, done approximately from 1740 to 1870, consisted of straightening and removal of small river islands (Schmidt 2000). Also dams were built, predominantly on the lower Elbe. Had that river training the potential to influence the occurrence of Elbe floods? Evidently, this question regards more the water stage and less the runoff.

To answer this question, Mudelsee et al. (2004) posited that effects of river training – if existing – should have been in the similar direction, regardless of the magnitude or the season. However, Figure 4.2 shows heterogeneous trends: upward for heavy winter floods and downward (but not significant) for heavy summer floods. The trends for minor floods (class 1) are indistinguishable from zero for the interval from the eighteenth to the nineteenth century. We conclude that river training had only a minimal influence on flood risk for the middle Elbe. Mudelsee et al. (2004) found minimal influences also for the neighboring river Oder.

4.1.4 Land Use

Collectivization of farmland in the area of the middle Elbe below Dresden changed agricultural land use since the foundation of the German Democratic Republic in 1949. Fields were increasingly cultivated with heavy equipment. This may have contributed to enhanced surface runoff in periods with large precipitation and increased flood risk (van der Ploeg and Schweigert 2001).

The application of the trend test to the runoff record from Elbe station Dresden for the interval from 1949 to 2013, however, does not support land-use changes as a driver of flood risk (Table 4.3).

Owing to the small sample sizes ($m = 15$ and 8 for class 1–3 floods in winter and summer, respectively), the resulting P-values are rather large, and the

Table 4.3. *Cox–Lewis test results on Elbe floods for the interval from 1949 to 2002. The results of the test (Section 3.2.5) of the null hypothesis "constant occurrence rate" (no trend), obtained on a sample of size m of extreme events, are given as test statistic (u_{CL}) and one-sided P-value.*

Season	Magnitude class	m	u_{CL}	P
Winter	1–3	15	1.04	0.15
Summer	1–3	8	−0.43	0.33

trends (upward for winter and downward for summer) are not significant. For class 2–3 events, the sample sizes are insufficient for a meaningful application of the test.

Mudelsee et al. (2004) evaluated another type of land-use change. Could deforestation have reduced the retention capability of the land and enhanced runoff in the mountainous catchment area (Erzgebirge) before the instrumental period? Bork et al. (1998) considered that the Millennium Flood in July 1342, which affected many parts of Germany, was related to deforestation. However, the existing evidence of deforestation in the Erzgebirge is not convincing (Mudelsee et al. 2004). Although one cannot definitively rule out deforestation, we prefer data inhomogeneity (Section 4.1.1) as explanation of the observed increases in flood risk before 1500 (Figure 3.9).

4.1.5 Climate

Data inhomogeneities such as document loss, especially for minor events, prevent the interpretation of flood occurrence estimations for the interval before AD 1500 (Section 4.1.1). For the interval thereafter, river training in the form of reservoir building (Section 4.1.2) and other cultural measures (Section 4.1.3) had only small effects on the occurrence rate of Elbe floods. Also the other anthropogenic influence – land use – had a minimal impact (Section 4.1.4). Left as a driver of flood risk over time is climate change. We have to keep in mind that the discussion is about flood occurrence rates for the middle Elbe and regional climate changes in the affected catchment area on decadal to centennial timescales.

Elbe winter flood risk peaked prominently in the second half of the sixteenth century (Figure 3.9c), where $\widehat{\lambda}(T) \approx 0.15$ a^{-1} (a heavy flood every sixth to seventh winter on average). Similar increases for that time interval were found also for central and southwest European rivers (Brázdil et al. 1999).

This fits into the picture, since the sixteenth century is considered as rich in precipitation, also in summer, see Mudelsee et al. (2004) and references cited therein.

Elbe winter floods then became less frequent, and a minimum with $\widehat{\lambda}(T) \approx$ 0.08 a^{-1} was approached at around 1700, which the 90% confidence band indicates as significant (Figure 3.9c). This reflects the dry and cold European climate, which may be related to the late Maunder Minimum, a period of few sunspots and reduced solar activity (Luterbacher et al. 2001). More detailed spatial agreement provides Starkel (2001), who noted the absence of heavy rains in the Polish/Czech area at around that time.

After the dryness, the risk of heavy Elbe winter floods increased again and reached a peak in the first half of the nineteenth century, when $\widehat{\lambda}(T) \approx 0.21$ a^{-1}. Since then, a significant downward trend prevailed, which is attested by the confidence band (Figure 3.9c) and the hypothesis test (Table 4.2).

What caused the long-term trend? Breaking ice at the end of winter may function as water barrier and enhance a high water stage severely. Ice floods, such as in February–March 1784 (Weikinn 2000), were ubiquitous for the Elbe in former centuries. The cause for the trend toward reduced winter flood risk was climate change in the form of regional warming (Mudelsee et al. 2003, 2004). Elevated winter temperatures in the region (Stocker et al. 2013) may have reduced flood risk in two ways: first, via a reduced rate of strong river freezing and related ice jam; second, via a reduced rate of occurrence of a frozen soil, which has a low absorbing capacity. Could increased salt concentrations in the river water have reduced the freezing point and acted as a cause, instead of regional warming? Mudelsee et al. (2004) estimated that the amount of salt in the river was insufficient.

4.2 Case Study: Monsoon Droughts

The Intertropical Convergence Zone (ITCZ) is a narrow latitudinal band of wind convergence and associated precipitation. The Tibetan plateau heats during boreal summer, and the resulting low-pressure regime further "sucks" the moist air flow over the Indian Ocean south of the ITCZ (Figure 4.5a). During winter, this pattern is reversed (Figure 4.5b). Such a changing wind and precipitation system is called monsoon (Webster et al. 1998). The Indian Ocean monsoonal rainfall undergoes changes also at longer timescales, caused by changes in solar insolation (Fleitmann et al. 2003).

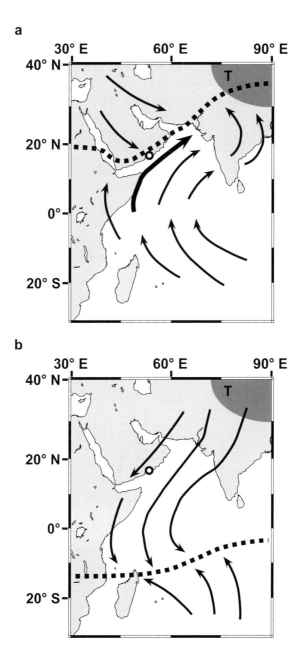

Figure 4.5 Map of the region of the Indian Ocean monsoon. The position of the ITCZ (dashed line) switches between boreal summer (a) and winter (b); in turn, also the predominant wind directions (arrows) change. The site of stalagmite Q5 (open circle) receives most of the precipitation from June to September, mostly via transport by the Somali Jet Stream (a, thick arrow). T, Tibetan plateau. Drawn after Fleitmann et al. (2007)

Figure 1.3 introduced a Holocene proxy record of Indian Ocean monsoonal rainfall. The data come from stalagmite Q5 from Qunf cave in Oman (Figure 4.5) and cannot be assumed to represent the full monsoon region. Here we analyze this dataset and focus on extreme droughts, which we first detect (Section 4.2.1) and then study in their rate of occurrence (Section 4.2.2).

4.2.1 Detection

Stalagmite Q5 in Oman began to grow at 10,300 a BP (Figure 4.6). This indicates that the monsoon, as documented by $\delta^{18}O$ from Q5, set in at that place at around that date. It then approached a wet phase in the earlier half of the Holocene, as reflected by the most negative $\delta^{18}O$ values. The reason is that local insolation – one driver of the monsoon – gradually decreased due to changes in the geometry of the Earth's orbit (Fleitmann et al. 2003). Somewhere between 6000 and 8000 a BP, the monsoon began to weaken. It got drier, and stalagmite Q5 ceased to grow at 2741 a BP (Figure 4.6).

In light of that prior physical knowledge, we ignore running-median estimation of the trend (Section 2.4) and employ instead a parametric description. The trend model comprises a ramp in the earlier part (up to the wet phase),

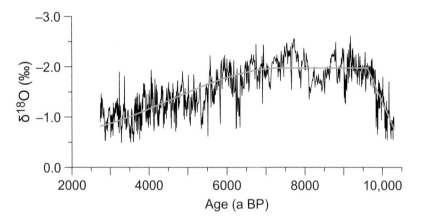

Figure 4.6 Indian Ocean monsoon rainfall during the Holocene: trend. A combination of a portion of a ramp (Section D.1.3) in the early part and a sinusoid in the late part is fitted via OLS as trend function (gray line) to the time series (black line). A sinusoid is employed because Holocene changes in local solar insolation, induced by Earth orbital changes (i.e., of a trigonometric functional form), influenced monsoonal rainfall amounts. For the numerical values of the trend parameters, see Section 4.2.1. Adapted from Mudelsee (2014: figure 4.18 therein)

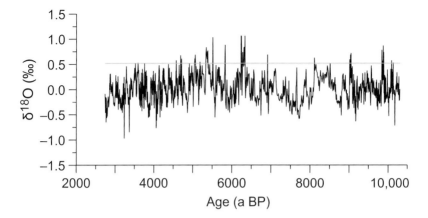

Figure 4.7 Indian Ocean monsoon rainfall during the Holocene: drought detection. The $\delta^{18}O$ anomalies (data minus trend) (black line) are plotted on a non-inverted vertical scale, that means, dry extremes are "upward." A constant threshold ($l = 672$, i.e., $2l + 1 = n = 1345$) (gray line) is used to detect POT extremes. A relatively low threshold (parameter $z = 3.0$) is utilized to allow a sufficient number of exceedances ($m = 38$).

and a sinusoid thereafter (Figure 4.6). The estimated change-point values are (10,300 a BP, $-0.77\%_0$), (9617 a BP, $-1.98\%_0$), and (7200 a BP, $-1.98\%_0$). For error bars and estimation details, see Mudelsee (2014: figure 4.18 therein).

Since low $\delta^{18}O$ values reflect strong monsoonal rainfall (Figure 1.3), the aim is to detect positive extremes as drought indicators. Since a time-dependent trend has already been subtracted and the $\delta^{18}O$ anomalies calculated, let us place a constant threshold (Figure 4.7). A relatively low threshold (parameter $z = 3.0$) yields a number of 38 exceedances.

Before proceeding with the estimation of the drought occurrence rate, however, an autocorrelation analysis is necessary to study the condition of independence of the exceedance times (Section 2.2). The persistence time (Eq. 2.8), estimated on the $\delta^{18}O$ anomalies using the software TAUEST (Mudelsee 2002), is $\hat{\tau} = 15.0$ a with 90% percentile confidence interval (CI) [13.1 a; 17.2 a]. Hence we say that if

1. an extreme event is separated from the next extreme by less than $\hat{\tau}$, and
2. there are no below-threshold data in between,

then both extremes belong to the same cluster of dependent extremes. The task is thus to determine the clusters and retain only the maximum of the extremes within. The number of independent extreme drought events is reduced by this exercise from $m = 38$ to $m = 26$ (Figure 4.8).

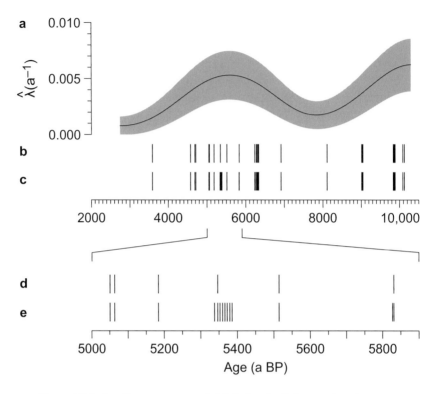

Figure 4.8 Indian Ocean monsoon rainfall during the Holocene: drought occurrence rate. The $m = 38$ POT exceedance times (c, e) are reduced to $m = 26$ independent events (b, d) by means of an autocorrelation analysis (Section 4.2.1). The effect of this procedure is illustrated for the interval [5000 a BP; 5900 a BP] (d, e). The occurrence rate (a, solid line) with 90% confidence band (shaded) is estimated by means of a Gaussian kernel with a bandwidth of $h = 800$ a and 2000 bootstrap resamples.

4.2.2 Occurrence Rate Estimation

Boundary bias correction via the reflection method (Section 3.2.3) is relevant for the right boundary (early part) since there sit a number of events (Figure 4.8c), while it is less so for the left boundary. The cross-validation function (Figure 4.9) shows a clearly expressed local minimum at around $h = 800$ a. We follow Marron (1988) and employ this value since it may reveal true structures in the estimation curve. An explanation for the even smaller cross-validation values for $h \to 100$ a may be that the autocorrelation correction (Section 4.2.1) did not completely remove all dependent events. A small h could try to reveal this spurious structure – which we avoid by selecting a larger bandwidth value.

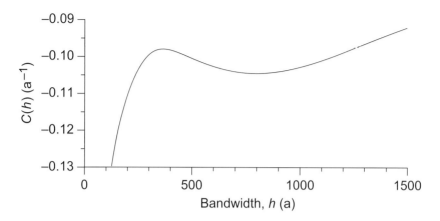

Figure 4.9 Indian Ocean monsoon rainfall during the Holocene: cross-validation function. The function $C(h)$ (Eq. 3.17) for the extreme droughts has a local minimum at $h \approx 800$ a.

The estimation curve (Figure 4.8a) shows that at the onset of the monsoon, the drought occurrence was at a maximum $(\widehat{\lambda}(T) \approx 0.006 \text{ a}^{-1})$. Then it decreased steadily and reached a low at around 7828 a BP at a level of approximately 30% of the maximum. Subsequently, a local maximum in drought risk was reached $(\widehat{\lambda}(T) \approx 0.005 \text{ a}^{-1}$ or 5 events per millennium on average at around 5585 a BP). A downward trend in drought occurrence prevailed until the end of the observation interval (Figure 4.8a).

4.3 Outlook

We have studied in this chapter on floods and droughts two cases, where we applied the methods (Chapter 3) to time series data from certain geographical sites. The scientifically natural approach to achieve a better, wider assessment of the results is to explore other study directions. There are three obvious directions.

First, space. This means to analyze data from other regions. Second, time. This means to extend the time range for the studied site. We can look back in time with the help of new measurements (Appendix A). It may also be done forward in time by means of more recent measurement data or, if we look into the future, by means of climate models (Appendix C). Third, resolution. This means to achieve a higher time resolution (i.e., smaller average spacing) for the same geographical site and time range. In the case of climate proxies, this task may consist in performing new measurements on the same climate archive, perhaps by employing an improved laboratory technique or a better measuring device.

The natural approach, also called normal science (Kuhn 1970), is what most climate researchers do most of their time. The other approach is to look on the same data but from another viewpoint, under another paradigm, that is, using another scientific language (e.g., hydrology instead of meteorology) and another set of analytical tools (e.g., indices such as those used for temperature extremes in Chapter 5). Such a paradigm shift (Kuhn 1970) may be necessary if the data interpretation under the old paradigm becomes problematic. Major shifts do not happen often.

The case study of the Elbe floods (Section 4.1) can be extended in the first, spatial dimension by studying neighboring rivers. The neighbor to the east, the river Oder, has been analyzed (Mudelsee et al. 2003, 2004) for the same time range (past millennium) and using the same seasonal differentiation and statistical methods as done for the Elbe. A smaller neighbor to the west, the river Werra, has also been studied (Mudelsee et al. 2006), however, for a shorter time range owing to limited availability of documentary data. The results show some agreement with those for the Elbe. For example, also the Oder reveals a downward trend in winter flood risk over the past century due to fewer strong freezing events – an effect of regional warming (Mudelsee et al. 2003). It becomes evident that the Clausius–Clapeyron equation, which quantifies how the atmosphere's water carrying capacity increases with temperature, is too simple for the explanation of the changes in the hydrological cycle that are associated with climate changes. More factors are involved. For example, the preferred wind directions and orographic rainfall intensities may also change (Mudelsee et al. 2004). A recent study on European rivers for the interval from 1960 to 2010 identified besides precipitation also soil moisture and spring temperatures (i.e., snow melt) as driving factors (Blöschl et al. 2019).

The extension of the documentary Elbe flood record back in time (second dimension) seems impossible due to lack of data. Even if such were present, the loss of data would exacerbate the inhomogeneity problem (Box 2.1). Natural archives, however, can serve to achieve the extension (Wilhelm et al. 2019). For example, Jílek et al. (1995) took sediment samples from the Holocene floodplain of the river Labe (how the river is called in the Czech Republic). They searched for and dated remnants of large tree trunks as flood indicators. One caveat is that the extended record represents the upper part of the river, while the documentary Elbe flood record the middle part. Another is that the proxy flood record bears no information about the hydrological season.

Also the increase of the monthly temporal resolution (third dimension) of the Elbe flood record seems rather difficult. The Weikinn source texts (Box 2.1) do indeed sometimes contain the exact date of an event and also the used calendar (Julian or Gregorian). However, too often only the month is given. Daily resolution can be achieved for the instrumental period (Figure 3.2), but for the best Elbe station, Dresden, this goes back "only" to 1806 and not the past thousand years.

The case study of Holocene monsoon droughts (Section 4.2) is regionally representative for the western Indian Ocean region (Figure 4.5). However, the monsoon can be described as a system that is forced – Asian-wide and beyond – by the seasonal migration of the ITCZ (Sinha et al. 2011). Thus, our case study can be extended spatially, and the extension to the East Asian monsoon region for the Holocene was achieved using the stalagmite climate archive from the Dongge cave (China) and $\delta^{18}O$ as rainfall indicator (Wang et al. 2005). Some of the weak East Asian monsoon intervals (Wang et al. 2005: figure 1 therein) do have counterparts in the western region (Figure 4.7), such as the drought at around 6.2–6.3 ka BP, some do not. More detailed comparison work of events and occurrence rates can be done.

Fleitmann et al. (2007: figure 1 therein) show a map of Holocene proxy archive sites, which includes also information about the time ranges and the average spacing of the time series. Such types of maps (updated to the present knowledge state) serve the other types of extension of the monsoon study directions. It seems fair to say that currently we climate researchers lack a sound understanding (empirical evidence and theory) of Holocene monsoon droughts.

The Palmer Drought Severity Index (PDSI) is a widely used indicator that can be calculated for a given site just from the temperature and the geographical latitude. However, this measure may be too simple, and the PDSI may be an inaccurate and biased drought indicator (Sheffield et al. 2012). That study employed a more realistic measure and found "little change in global drought over the past 60 years" (Sheffield et al. 2012: p. 435 therein). However, this global view is likely less relevant for California (Seager et al. 2015; Swain et al. 2018) and Australia (Biswas and Mosley 2019), which both were hit by recent prolonged droughts. Precipitation, evaporation, and droughts may show strong spatial variability. More regional studies based on high-resolution observations and climate model output are required for achieving a better understanding of drought risk under climate change and the consequences for ecosystems such as forests (Anderegg et al. 2018).

Evidently, other types of extremes associated with water can be studied that are not so directly related with precipitation, such as tsunamis (Rubin et al. 2017) or coastal flooding caused by rising sea levels (Witze 2018).

4.4 Summary for the Risk Analyst

Economically relevant are the current situation and the near future (the next five years). For strategic decisions, also the next few decades count. The ideal resolution of input data for this time range is daily or better (i.e., smaller). Precipitation and other variables related to the water cycle show large spatial variations. Data from a single site therefore have a limited geographical representativeness, and many sites have to be studied to achieve a comprehensive analysis of a larger region.

River flood risk, in Europe and elsewhere, is not well understood. It becomes evident that the Clausius–Clapeyron equation is too simple for the explanation of the changes in the hydrological cycle that are associated with climate changes. More factors are involved. Depending on the location, these may include (1) wind directions and orographic rainfall, (2) soil moisture, and (3) spring temperatures.

Good-quality, homogeneous river data facilitate the analysis. Sensitivity analyses permit to inspect more details, for example, to study the effects of (1) seasons (hydrological summer and winter), (2) river training, and (3) land-use changes.

Such data pretreatments allow also to study multivariate extremes and cascading events. Consider, for example, joint extremes of river floods of the Elbe and the neighboring river Oder (Figure 4.1). The detection (Section 2.4) can straightforwardly be adapted such to require both rivers to show (within a time corridor) an extreme. Such analysis types are not restricted to river floods. Consider the joint occurrence of floods and winds. For example, the impacts of the elevated water levels of the river Mississippi in July 2019 were exacerbated by the presence of hurricane Barry, which made landfall in Louisiana in the United States. Or consider the Fukushima nuclear disaster on 11 March 2011, where a tsunami interfered with other types of extremes in the energy sector.

Droughts appear somewhat better understood than river floods since they are also determined by temperature extremes, and temperature shows smaller spatial variations than precipitation (Chapter 1). On the other hand, the duration aspect (i.e., how to define a drought) introduces new uncertainties. This aspect is studied further in the context of heatwaves (Chapter 5).

The duration aspect is less relevant for the case study of Holocene monsoon droughts (Section 4.2) since the proxy record has a relatively large time resolution.

The stationarity assumption (i.e., time-constant occurrence rate) is almost obsolete in this world under climate change. At least, it has to be tested using methods from Section 3.2.5. And if you do fit a stationary process with a certain distribution, then adopt flexible three-parametric forms such as the GEV (Section 3.1.1) or the GP (Section 3.1.2) and ignore recommendations from traditional hydrologists to adopt the two-parametric Gumbel distribution.

Under nonstationarity, the task of predicting the time-dependent occurrence rate (e.g., Figure 3.9) into the near future is likely best performed by means of extrapolation of the curve. Due to the lack of knowledge about the driving factors, flood risk prediction by means of coupling hydrological models to climate models – is not yet feasible. These tools are useful for the study of what-if questions and strategic development, not for risk quantification.

Finally, a hint on analytical methodology. The index approach to study temperatures extremes can also be applied to precipitation series, see Section 5.1 and the literature in the Reading Material for that chapter.

Reading Material

Mudelsee et al. (2003) reconstruct Elbe and Oder floods over the past millennium, estimate occurrence rates, and attribute causes. Mudelsee et al. (2004) study meteorological causes of floods and provide details for Mudelsee et al. (2003). Munoz et al. (2018) extend the instrumental flood record of the river Mississippi back to AD 1500 by means of proxy data. Silva (2017) reviews the nonstationary analysis of hydrological variables. Wilhelm et al. (2019) review the extension of instrumental flood records back in time by means of documentary and natural climate archives.

Webster et al. (1998) is a review of monsoons. Fleitmann et al. (2007) review proxy techniques for the reconstruction of past states of the Indian Ocean monsoon.

Mudelsee (2002) introduces persistence time estimation.

Kuhn (1970) coined the expressions "paradigm shift" and "normal science" for the philosophy of science.

5

Heatwaves and Cold Spells

Extremely high temperatures are associated with heatwaves, and extremely low temperatures with cold spells. However, there is also the duration aspect: longer exceedances of a high threshold have a stronger impact on natural systems or society (e.g., human health) than shorter exceedances. There exist several ways to define a heatwave or a cold spell by means of indices, which are calculated from the temperature and duration values. Our review (Section 5.1) finds not all indices useful – and it introduces two new ones.

The new indices, and some old, are then applied in case studies to instrumental temperature records. The first is the reanalysis of European heatwaves, such as the one in summer 2003 (Section 5.2). The second study is on cold spells and late frosts at Potsdam (Section 5.3). We indicate current research directions (Section 5.4). Finally, we give a short summary for risk analysts and other people who wish to quantify current and near-future trends in the occurrence of heatwaves and cold spells (Section 5.5).

5.1 Indices

It is reasonable that several heatwave definitions are in usage since this type of climate extreme influences various sectors (health, agriculture, etc.) and different geographical regions. Another reason may be the reduced willingness of researchers to look beyond the boundaries of their scientific disciplines. Zhang et al. (2011), Hartmann et al. (2013: box 2.4 therein), and Perkins and Alexander (2013) offer comprehensive overviews and list many indices. Following these references, a useful index for warm or cold temperature extremes should meet several requirements.

Requirement 1   The index should be applicable across a wide range of sectors. It should capture both aspects of an extreme: magnitude and duration. The concept of its calculation should be accessible to researchers, technicians, and end users.

Requirement 2   The index should be applicable to a wide range of geographical regions.

Requirement 3   Data have to be available (e.g., daily temperature values) for a meaningful index calculation. The number of missing data points should be small, but it may be greater than zero.

Requirement 4   The definition of the index should be clear and concise in order to permit others to repeat the calculations and reproduce the results (given the data). It may be necessary to introduce some parameters for the definition, such as a threshold. However, the fewer of such adjustable parameters, the better – Ockham's Razor.

Box 5.1   **Personal Reflection: Ockham's Razor and the philosophy of parsimony**

Consider a time series sample comprising just two points. A perfect straight line can be fitted to the data. The fit is described by two parameters, slope and intercept. Next, consider a series with $n = 3$ points. A perfect parabola (i.e., a polynomial of order two) can be fitted to the data. The fit is described by three parameters. Now, consider the time series of the annual mean of global surface-air temperature for the interval $[1880; 2017]$ with $n = 138$ points (Mudelsee 2019). A perfect polynomial of order 137 can be fitted to the data. How do you interpret the 138 parameter estimates? Here you are lost. However, you can also fit a straight line to the $n = 138$ data points (Section D.1.2). The fit is not perfect. The parameter estimates are uncertain. You find out that the slope estimate is clearly larger than its estimation uncertainty. That means, you have quantified global warming, you have quantified with error bar how much warmer it gets per decade. To summarize, by keeping the model simple (straight line) with regard to the data, you learned something useful.

   "Keep things simple!"

   This is the heuristic imperative we followed in the temperature example. It is commonly associated with the name of William of Ockham (circa 1285–1347). Ockham's Razor cuts away the unnecessary

$138 - 2 = 136$ parameters in the temperature example. Ockham's Razor, more generally, removes the metaphysical stuff, the statements that cannot be falsified and that belong therefore not to science (Popper 1959).

As often in the philosophy of science, the expression "Ockham's Razor" has an intricate history. Hübener (1983) reports the following. The term was coined in 1649 by Libertus Fromondus (1587–1653). The heuristic imperative goes back to Aristotle (384 BC–322 BC), who wrote, for example, in his "Physics" (Book I, 188[a]17–18) (Aristotle 1936: p. 343 therein): "It is better to make the elements fewer, i.e., finite in number, as Empedocles does." William of Ockham himself did not so strictly use his tool. However, Russell (1996: p. 465 therein) praised his merit: "By insisting on the possibility of studying logic and human knowledge without reference to metaphysics and theology, Occam's work encouraged scientific research."

Also Isaac Newton (1643–1727) and Gottfried Wilhelm Leibniz (1646–1716) are said to have followed the stronger form of the imperative. Newton (1687: p. 402 therein) wrote: "We are to admit no more causes of natural things than such as are both true and sufficient to explain their appearances." In short, this means:

"Keep things simple. But not too simple!"

Back to the temperature example. The linear model may be too simple. If you wish to examine whether there has been an acceleration or a pause of global warming (Mudelsee 2019), then you have to invoke change-point models (Section D.1.3), that means, you have to invest more parameters than just two.

Within the context of this book – climate change and climate extremes – Ockham's Razor teaches us two lessons. First, a parametric description of the distributional shape (Section 3.1) should not only include location and scale, but also a third parameter, shape. Second, the stationarity assumption may in many analysis situations be obsolete because of climate change. We have to invest more parameters to model the time dependence. "Sufficient. Not too simple."

Although many of the indices from Tables 5.1 and 5.2 are commonly employed, some miss to a certain degree the mentioned requirements. Hence, we consider not all of the indices useful for the identification and statistical analysis of temperature extremes.

Table 5.1. *Indices for heatwaves. The list is a selection from the Fifth Assessment Report of the IPCC (Hartmann et al. 2013: box 2.4 therein), augmented by the index "WSDI" (Zhang et al. 2011); also shown are the new "action measures." For TX90p and TN90p, also relative units are employed (percentage of season/year instead of days). These two indices are calculated with calendar day percentiles. We also consider absolute percentile values (TX|90p| and TN|90p|). Stronger extremes, such as TX95p based on the 95th percentile, are also used.* T_{max}, *maximum daily temperature;* T_{min}, *minimum daily temperature.*

Index	Description	Definition	Unit
TXx	Warmest T_{max}	Seasonal/annual maximum of T_{max}	°C
TNx	Warmest T_{min}	Seasonal/annual maximum of T_{min}	°C
TX90p	Warm days	Seasonal/annual count of days when T_{max} > calendar day 90th percentile	d
TN90p	Warm nights	Seasonal/annual count of days when T_{min} > calendar day 90th percentile	d
TR	Tropical nights	Seasonal/annual count of days when T_{min} > 20 °C	d
WSDI	Warm spell duration index	Seasonal/annual count of days when T_{max} > 90th percentile on ≥ 6 consecutive days	d
ATX|90p|	Action measure for warm days	Integral of exceedance (T_{max} − 90th percentile) over duration on ≥ 3 consecutive days	°C · d
ATN|90p|	Action measure for warm nights	Integral of exceedance (T_{min} − 90th percentile) over duration on ≥ 3 consecutive days	°C · d

The maximum daily surface-air temperature at Orléans (Figure 5.1) serves to illustrate some heatwave indices (Figure 5.2). Let us first consider which indices to avoid in the analysis of temperature extremes.

The adoption of the seasonal/annual extreme, TXx, has the problem that the duration aspect (Requirement 1) is not taken into account. This applies also to the other extremes (TNx, TXn, and TNn). For Orléans during the JJA season in 2003 (Figure 5.2a), the value is TXx = 39.9 °C.

The usage of seasonal/annual count of days of threshold exceedances does take into account the duration. This applies also to cold events (negative exceedances). The adoption of absolute threshold values, such as 20 °C for TR, introduces the problem that the results for different regions (e.g., Antarctica versus Sahara) may become difficult to compare. This means a

Table 5.2. *Indices for cold spells. The list is a selection from the Fifth Assessment Report of the IPCC (Hartmann et al. 2013: box 2.4 therein), augmented by the index "CSDI"(Zhang et al. 2011); also shown are the new "action measures" and a new indicator of late frosts. For TX10p and TN10p, also relative units are employed (percentage of season/year instead of days). These two indices are calculated with calendar day percentiles. One may also consider absolute percentile values (TX|10p| and TN|10p|). The calculation of the Last Frost Day (LFD) and of the Start of Growing Season (SGS) may impose additional conditions on the allowed days in a year,which further may depend on the geographical location.* T_{max}, *maximum daily temperature;* T_{mean}, *mean daily temperature;* T_{min}, *minimum daily temperature.*

Index	Description	Definition	Unit
TXn	Coldest T_{max}	Seasonal/annual minimum of T_{max}	°C
TNn	Coldest T_{min}	Seasonal/annual minimum of T_{min}	°C
TX10p	Cold days	Seasonal/annual count of days when T_{max} < calendar day 10th percentile	d
TN10p	Cold nights	Seasonal/annual count of days when T_{min} < calendar day 10th percentile	d
FD	Frost days	Seasonal/annual count of days when T_{min} < 0 °C	d
CSDI	Cold spell duration index	Seasonal/annual count of days when T_{min} < 10th percentile on ≥ 6 consecutive days	d
ATX|10p|	Action measure for cold days	Integral of negative exceedance (10th percentile − T_{max}) over duration on ≥ 3 consecutive days	°C · d
ATN|10p|	Action measure for cold nights	Integral of negative exceedance (10th percentile − T_{min}) over duration on ≥ 3 consecutive days	°C · d
LFSI	Late Frost Severity Index	LFSI = LFD − SGS, where LFD is the last day in a growing season with T_{min} < 0 °C and SGS is the first occurrence of six consecutive days with T_{mean} > 5 °C.	d

violation of Requirement 2. However, frost (FD) offers a wider comparability of cold events.

It is therefore common practice to replace absolute thresholds by percentiles of the empirical distribution. The advice (Folland et al. 1999) to fit an extreme value distribution (Sections 3.1.1 and 3.1.2) for percentile estimation is less often followed in practice. For Orléans during the JJA season (Figure 5.1b), which has $n = 6164$ values over the full period 1946–2012, the 90th percentile

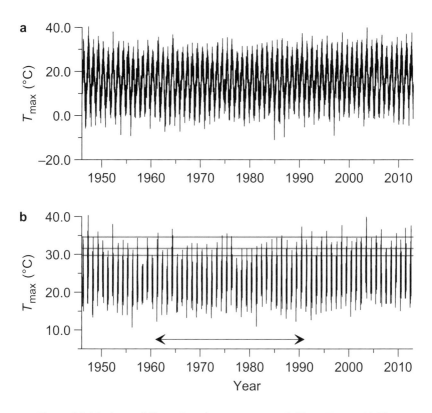

Figure 5.1 Maximum daily surface-air temperature at Orléans, France. (a) The shown interval is from 1 April 1946 to 31 December 2012. (The full instrumental series starts earlier but has many data missing.) (b) Extracted June–July–August (JJA) series; the data size is 6164. Four missing values (17 July 2007, 2 August 2008, 16 June 2009, and 3 August 2010) are inserted as the midpoints of the two neighboring days, respectively. This does not influence the estimation of the 90th, 95th, and 99th percentiles (gray horizontal lines), which are clearly larger than the inserted values. Also shown is a typical base period, from 1961 to 1990 (double arrow), which is also regularly employed for percentile estimation. Data from https://wol-prod-cdn.literatumonline.com/pb-assets/hub-assets/rss/Datasets/_1_C1121Winter-1513681076397.zip (7 February 2019)

is empirically estimated as the $[NINT(0.9 \cdot n) = NINT(5547.6) = 5548]$th largest value or 29.7 °C. (Interpolation between the 5547th and 5548th value, also possible, would give the same result.) The 95th and 99th percentile is estimated analogously as 31.6 °C and 34.6 °C, respectively (Figure 5.1b).

The selection of the percentile level is a trade-off problem (Section 2.4.1). On the one hand, many researchers, such as Perkins and Alexander (2013), see

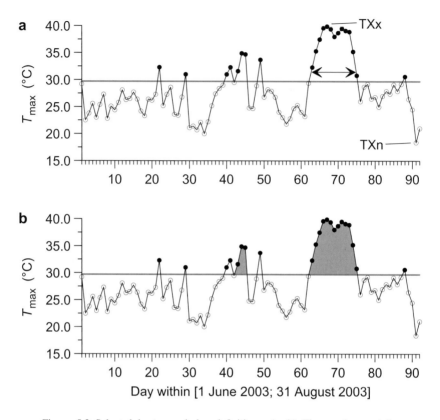

Figure 5.2 Selected heatwave index definitions. (a, b) The maximum daily surface-air temperature at Orléans (Figure 5.1) is exemplarily shown (open/filled symbols connected with a line) for the JJA season in 2003 with the 90th percentile (gray horizontal line) as threshold. (a) The indicated warmest and coldest T_{max} define for the year 2003 the TXx and TXn indices, respectively. The number of threshold exceedances (filled symbols) defines the TX|90p| index. (Note that the conventional definition employs the calendar day percentile and denotes the index as TX90p.) The count of threshold exceedances on six or more days (double arrow) defines the WSDI index. (b) The integral of the threshold exceedance curve over duration on three or more days (shaded areas) defines the ATX|90p| index. There are two such events in JJA 2003.

the 90th percentile as a good compromise (liberal versus conservative). On the other, in an explorative analysis it is important to play with the parameters and try also other settings.

It is important to note that the percentile estimation depends on the selected period in the presence of a long-term temperature trend – which is not unrealistic in a climate-change world. This may limit the comparability of results from different stations with different full time periods (Requirement 2). For example, the Orléans series covers 1946–2012, while the Potsdam series (Figure 1.2) covers 1893–2018. Usage of a common base period, such as 1961–1990 (Figure 5.1b), may therefore improve the comparability. However, there are two caveats. First, the base period is shorter than the full period, which means fewer data and larger estimation uncertainties. Second, the act of placing a threshold on basis of data "inside" (the base period) may introduce a bias in the occurrence rate of threshold exceedances "outside" (Zhang et al. 2005); this paper gives a correction for the bias problem in the form of computer simulations. The description of the procedure (by other authors) sometimes lacks accessibility (Requirement 1) or conciseness (Requirement 4). Let us stress that it is also important to play with the threshold parameter and study the sensitivity of results in order to obtain a robust assessment.

Irrespective of whether a percentile-threshold is calculated with or without a reference to a base period, there is the option to employ calendar day percentiles. This means to study not absolute temperature values but the anomalies that remain after the annual cycle has been removed. The idea is similar to that behind detection of extremes by means of a time-dependent threshold (Section 2.4.1), namely to allow for background and climate variability changes, in this case on a timescale of exactly one year.

The calculation of the anomalies is better done via the day-wise subtraction of day-wise averages than by removal of the annual peak in a Fourier spectrum (trigonometric functions) because the annual cycle needs not exactly follow a sinusoidal form (Folland et al. 1999). For Orléans and a base period of 1961–1990, this means to calculate the temperature average from the values for 1 June 1961, 1 June 1962 until 1 June 1990, subtract the average from all entries for the 1st of June and then to repeat this exercise for the 2nd of June until the 31st of August. Since for a 30-year base period such a mean calculation would be based on only 30 values, it is practice to employ five-day windows (Perkins and Alexander 2013). For example, the calculation of the average for the 1 June uses the entries for 30 May, 31 May, 1 June, 2 June, and 3 June – and it is then based on 150 values.

The conventional way to define warm days (TX90p) is as the seasonal/annual count of days when the calendar day 90th percentile is exceeded (Table 5.1). However, this focus on the anomalies misses aspects of the absolute values. For example, an extremely warm 1 June anomaly (e.g., the 1 June 1981 at Orléans with +5.61 °C) may have an absolute value (26.6 °C) that is not very relevant for human health. It is therefore important to consider also the absolute values.

We denote the indices calculated with calendar day percentiles (TX90p, TN90p, TX10p, and TN10p) following the convention (Tables 5.1 and 5.2), and we denote the indices calculated with absolute percentile values using mathematical absolute value symbols (TX|90p|, TN|90p|, TX|10p|, and TN|10p|). For Orléans during the JJA season in 2003 (Figure 5.2a), the warm days index is TX|90p| = 22 d; if relative units are employed, then TX|90p| = (22 d)/(92 d) = 23.9%.

One deficit of the warm days indices (TX90p or TX|90p|) is that they count also isolated days, that is, when just a single threshold exceedance occurred and no event before or after that day. This misses the duration aspect to some degree (Requirement 1). The WSDI overcomes this by imposing the condition that the exceedance has to occur on six or more days. For Orléans during the JJA season in 2003 (Figure 5.2a), the estimate of WSDI = 13 d refers to the exceptional interval from 2 to 14 August and ignores shorter earlier exceedances. The price to pay when adopting WSDI is that besides the treshold, a new duration parameter (which has a value of 6 d) is invoked. This brings additional arbitrariness and means a violation of Requirement 4 to some degree.

One drawback of the duration-based index WSDI is that the size of the threshold exceedances, that is, the magnitude aspect (Requirement 1), is not taken into account. Such a deficit has also the measure for cold spells (CSDI). One strategic line for the statistical analysis is to vary the threshold and study the sensitivity of the obtained WSDI results. Another line is to formulate a new measure that combines the magnitude with the duration aspect.

We call this new index type "action measure" in analogy to the concept of action in physics (where it is defined as the product of energy and time). For climate extremes analysis, we define it as the integral of the threshold exceedance curve over duration on three or more days (Figure 5.2b). The action measure has correspondences in cryology (Vaughan et al. 2013) and phenology (Flato et al. 2013), where the degree-day index quantifies the impact on ice sheets and plant growth, respectively. Also the new action indices invoke a duration parameter (which has a value of 3 d). This conflict with Requirement 4 seems inevitable: to take into account magnitude and duration, at least two adjustable parameters are required. Therefore, sensitivity analyses for both aspects are mandatory. One advantage of the action measures (e.g., of ATX|90p| versus TX|90p|) is that they allow for multiple events per season. For Orléans during the JJA season in 2003 (Figure 5.2b), the action measures for warm days are 12.1 °C · d for the earlier event (13–15 July) and 98.1 °C · d for the later, exceptional event (2–14 August). The associated event times are the interval centers (14 July and 8 August). Thus, the action indices have the POT data format (Section 2.2).

Finally, there exist heatwave definitions that include not only temperature but also relative humidity, such as the "apparent temperature" (Ta) (Fischer and Schär 2010). Such an index may be helpful for assessing heatwave impacts on human health – if humidity data exist. In line with Perkins and Alexander (2013), we will not consider Ta here because its calculation may generate conflicts with Requirement 3 (data availability) and Requirement 4 (parsimony in the formulation of the index).

5.2 Case Study: European Heatwaves

Europe is the home for about 750 million people (year 2019). An exceptionally strong heatwave occurred in the height of the summer 2003. The impact of the heatwave on human health is difficult to quantify since this was not a controlled experiment, which can be repeated. However, a serious attempt to estimate the excess mortality found a value in the order of fifteen thousand lives alone for France (Fouillet et al. 2006). Temperature is a climate variable closely related to atmospheric GHG concentrations (Chapter 1). Other variables, such as precipitation or wind speed, are "farther away." Thus, there is a direct concern that the rising GHG concentrations lead to rising occurrences of heatwave events.

We test the hypothesis of upward trends in the occurrence of European heatwaves by means of analysis of maximum daily surface-air temperature series from two stations, Orléans (Figure 5.1) and Potsdam (Figure 1.2). The analysis focus is robustness of results, therefore we play with the heatwave detection indices and study the sensitivity of results. A number of two stations is certainly not sufficient to capture fully the spatial aspects of heatwaves. Evidently, the case study can be extended to include many more stations in order to shed more light on the risk of European heatwaves.

5.2.1 Action Measures

The calculation of the action measures (Table 5.1) for European heatwaves requires to set two detection parameters: upper threshold and minimum duration. The sensitivity plots for JJA heatwaves at Orléans (Figures 5.3–5.5) show for all parameter settings that the heatwave from 2 to 14 August 2003 (Figure 5.2) is the strongest event recorded (since 1946). Hence, this is a robust finding. The plots further reveal that too high thresholds (99th percentile)

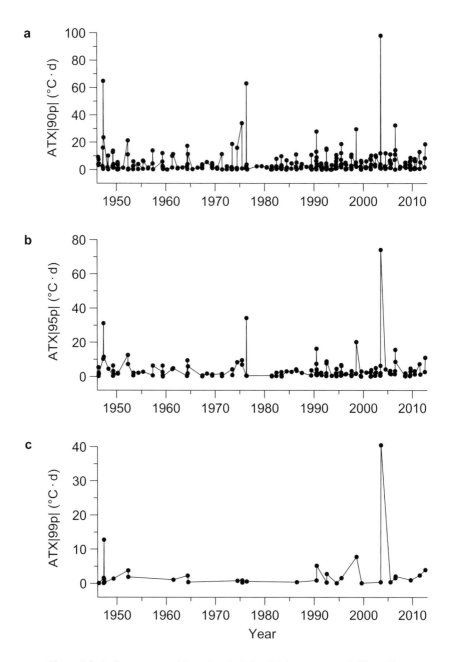

Figure 5.3 Action measures (short duration) for JJA heatwaves at Orléans. The index calculation adopts a duration parameter of one day, absolute percentile thresholds (a, 90th, 29.7 °C; b, 95th, 31.6 °C; c, 99th, 34.6 °C) for maximum daily temperature, and no base period.

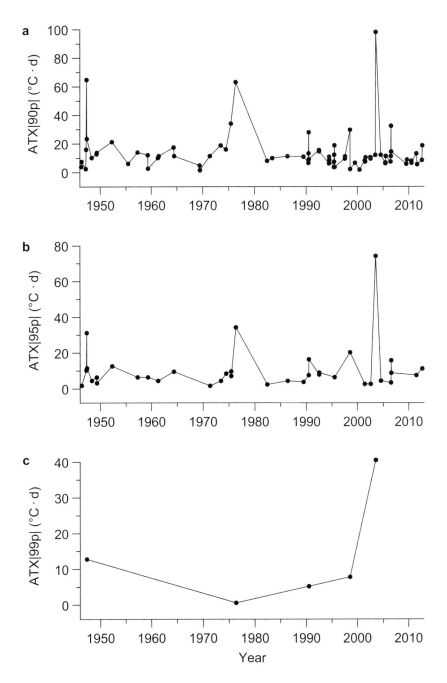

Figure 5.4 Action measures (medium duration) for JJA heatwaves at Orléans. As in Figure 5.3, but with a duration parameter of three days.

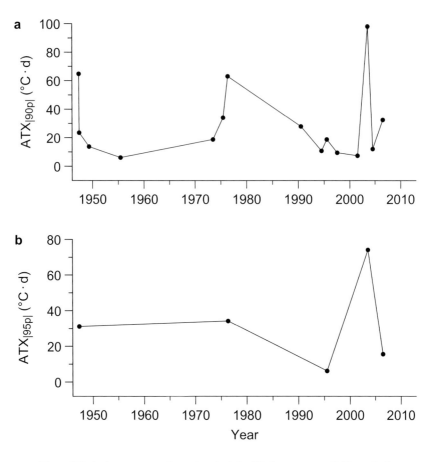

Figure 5.5 Action measures (long duration) for JJA heatwaves at Orléans. As in Figure 5.3, but with a duration parameter of six days. The ATX|99p| curve (not shown) has only one event.

or too long minimum durations (six days) are too strong conditions, which lead to rather few detected extremes and large estimation uncertainties. On the other hand, a minimum duration of one day (Figure 5.3) means basically that the duration aspect is not taken into account. We avoid also this too liberal condition. An adequate parameter setting is achieved with a minimum duration of three days and the 90th or 95th percentiles (Figure 5.4a, b).

   The sensitivity plots for JJA heatwaves at Potsdam (Figures 5.6–5.8) reveal that the biggest event was not in August 2003 (ATX|90p| = 8.6 °C · d) but rather centered around 27 July 1994 (ATX|90p| = 50.2 °C · d). Also for Potsdam, an adequate heatwave detection is achieved with a minimum duration of three days and the 90th or 95th percentiles (Figure 5.7a, b).

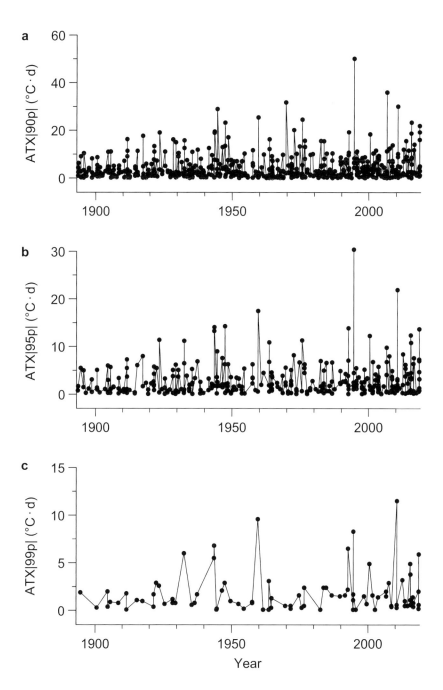

Figure 5.6 Action measures (short duration) for JJA heatwaves at Potsdam. The index calculation adopts a duration parameter of one day, absolute percentile thresholds (a, 90th, 29.5 °C; b, 95th, 31.3 °C; c, 99th, 33.9 °C) for maximum daily temperature, and no base period.

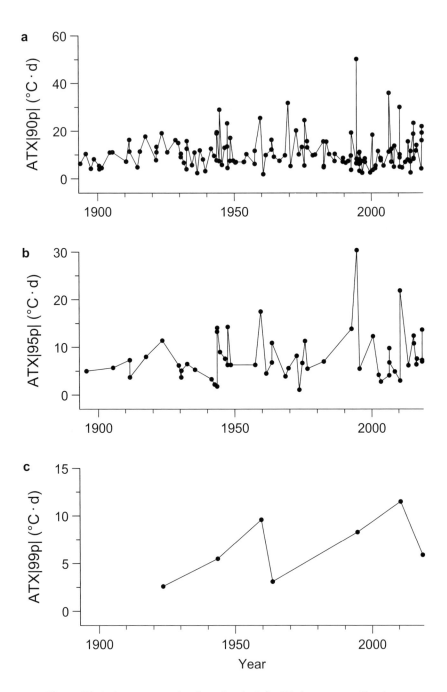

Figure 5.7 Action measures (medium duration) for JJA heatwaves at Potsdam. As in Figure 5.6, but with a duration parameter of three days.

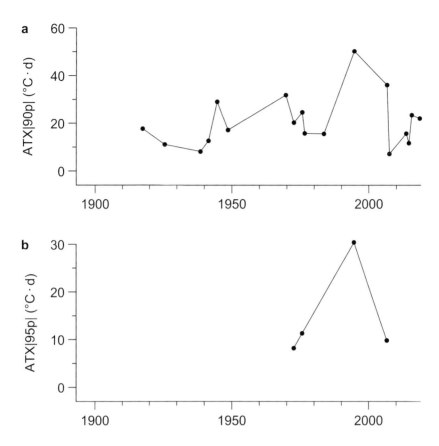

Figure 5.8 Action measures (long duration) for JJA heatwaves at Potsdam. As in Figure 5.6, but with a duration parameter of six days. The ATX|99p| curve (not shown) has no event.

The action measures successfully combine the threshold and duration aspects into one single impact number. In principle, also the sizes of the action measures can be compared with each other. The heatwave event at Orléans (Figure 5.4a, b), which has a maximum of ATX|90p| = 98.1 °C · d, seems to have been "more extreme" than all the events at Potsdam (Figure 5.7a, b), which have a maximum of ATX|90p| = 50.2 °C · d.

Such an initial conjecture can be further tested by means of estimation of a heavy-tail distribution (Section 3.1.4) to see whether Orléans has a smaller index estimate than Potsdam. The estimation is performed by means of the Hill estimator, a brute-force order selector, and 100 parametric Monte Carlo simulations. The results, however, are indistinguishable (Orléans:

$\widehat{\alpha} = 0.35 \pm 0.37$, Potsdam: $\widehat{\alpha} = 0.33 \pm 0.23$), which is due to the rather small sample sizes of ATX|90p| events (Orléans: $m = 72$, Potsdam: $m = 140$). To summarize, Orléans had a stronger heatwave impact measure than Potsdam, but it is not clear whether this is caused by a "heavier tail."

5.2.2 Occurrence Rate Estimation

The action index is in the form of POT data.This means that the event times when a heatwave occurred (in terms of ATX|90p| or ATX|95p|) can be further examined in a nonstationary analysis framework. We take a simple look and do not further differentiate between stronger and weaker values of the action measure. All event times for a station (shown as vertical bars in Figures 5.9 or 5.10) are jointly analyzed.

Boundary bias correction via the reflection method (Section 3.2.3) is relevant both for Orléans and Potsdam heatwaves since there are events located close to the lower and upper boundaries of the observation intervals. The cross-validation functions (not shown) indicate usage of a kernel bandwidth of $h = 5$ a. This value is set for all analyses (Figures 5.9 and 5.10) to facilitate the comparison of occurrence rate curves. The number of resamples ($B = 2000$) follows the usual recommendation (Section 3.2.4).

Although the JJA heatwaves at Orléans appear to have been "more extreme" than the events at Potsdam, the occurrence rate curves bear strong similarities. Let us first consider Potsdam since this is the longer series.

There is a local high in heatwave occurrence at Potsdam during the 1940s, where $\widehat{\lambda}_{\text{ATX}|90p|}(T) \approx 1.5 \text{ a}^{-1}$, that means, there have been on average one or two weaker events (90th percentile) per JJA season (Figure 5.10a). Orléans reports a similar $\widehat{\lambda}_{\text{ATX}|90p|}(T)$ estimate (Figure 5.9a). However, this series starts in 1946, and the estimate is influenced by the boundary bias correction. Also the stronger events (95th percentiles) reveal local highs during the 1940s for both stations (Figures 5.9b and 5.10b).

Potsdam shows for the decades before the 1940s clearly smaller summer heatwave occurrence rates, reduced by a factor of circa two (Figure 5.10). This is a robust finding owing to the excellent homogeneity of data from Potsdam – not only for wind speed (Section 2.2.1), but also for daily temperature.

After the 1940s with the elevated heatwave risk, the curves decreased over the following decades up to roughly the 1970s. Thereafter, both Orléans and Potsdam reveal upward trends. Depending on the threshold setting (90th or 95th percentile), these increases were moderate (Figure 5.9b) or strong (Figures 5.9a and 5.10).

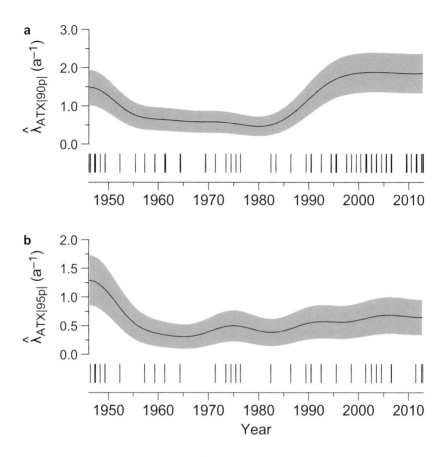

Figure 5.9 European heatwaves during the instrumental period: occurrence rates for Orléans. The detection index is the action measure, which is calculated with absolute percentile thresholds (a, 90th; b, 95th) and a duration parameter of three days (Figure 5.4a, b). The number of events is (a) $m = 72$ and (b) $m = 37$. The occurrence rates (solid lines) with 90% confidence bands (shaded) are estimated by means of a Gaussian kernel with a bandwidth of $h = 5$ a and 2000 bootstrap resamples.

Whereas for Orléans we observe a plateau behavior during the recent one or two decades (Figure 5.9), for Potsdam we note that the upward trends prevail to the present (Figure 5.10). However, the Potsdam series continues up to the 2018 summer season, while the Orléans series goes only to 2012. Consequently, the question after prevailing increases in JJA heatwave occurrence is difficult to answer for Orléans. Let us therefore apply the one-sided Cox–Lewis test (Section 3.2.5) for upward trends in JJA heatwave

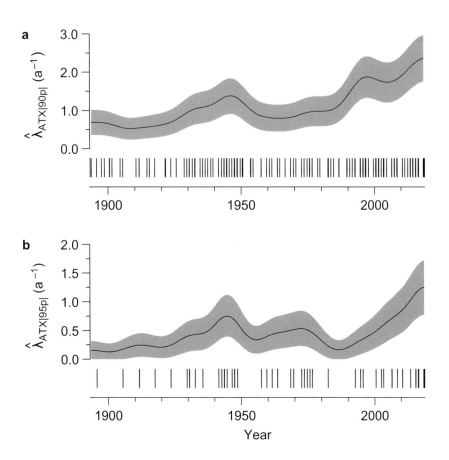

Figure 5.10 European heatwaves during the instrumental period: occurrence rates for Potsdam. The detection index is the action measure, which is calculated with absolute percentile thresholds (a, 90th; b, 95th) and a duration parameter of three days (Figure 5.7a, b). The number of events is (a) $m = 140$ and (b) $m = 54$. The occurrence rates (solid lines) with 90% confidence bands (shaded) are estimated by means of a Gaussian kernel with a bandwidth of $h = 5$ a and 2000 bootstrap resamples.

occurrence to the extreme event datasets, which are adapted such that the lower interval bound is set to 1 June 1980. The idea behind the adaptation is to inspect the recent decades. For Potsdam, the test yields highly significant upward trends. In the case of $\widehat{\lambda}_{ATX|90p|}(T)$, the number of events is $m = 65$, the test statistic is $u_{CL} = 2.00$, and the P-value equals 0.02; in the case of $\widehat{\lambda}_{ATX|95p|}(T)$, $m = 21$, $u_{CL} = 3.24$, and $P = 0.0006$. Also for Orléans and the smaller threshold (90th percentile), the increase is significant: $m = 47$, $u_{CL} = 2.49$, and $P = 0.006$. Only for the larger threshold (95th percentile), the result is not so strong ($u_{CL} = 0.95$ and $P = 0.17$), which may be due to the comparably small number of events ($m = 18$) recorded at Orléans. Therefore, the $\widehat{\lambda}_{ATX|95p|}(T)$ curve for Orléans (Figure 5.9b) should be assessed with more caution.

To summarize, we find for both stations occurrence rate plateaus or recent increasing trends, which lead to risk estimates for the present that are likely in excess of what has been observed before.

5.2.3 Other Measures

The analysis of the same temperature datasets by means of other indices may shed some light on the JJA heatwaves at Orléans and Potsdam. The WSDI (Table 5.1) is an index that, as the action measure, takes the duration aspect into account. To achieve robustness, we play with the WSDI detection parameters. The conventional setting is the 90th percentile as upper detection threshold and six days as minimum duration. We study also the 95th percentile and three days minimum duration.

The results for both Orléans (Figure 5.11) and Potsdam (Figure 5.12) reveal that the setting of the 95th percentile combined with six days minimum duration is too strict. For most of the years, there are no heatwave events that pass this criterion, and the WSDI series (Figures 5.11c and 5.12c) consist mostly of zeroes. It is difficult to learn from such series. Also the combination of the 90th percentile and six days minimum duration, that is, the conventional setting, shows this deficit (Figures 5.11a and 5.12a).

We learn more from WSDI in this case study when we relax the duration criterion. Setting the minimum duration as three days leads to many nonzero WSDI values distributed over the respective time intervals (Figures 5.11b, d, and 5.12b, d). This allows to learn about time-dependent trends in WSDI by means of regression analysis.

Straight lines fitted to the WSDI series (with three days minimum duration) reveal significantly increasing trends. For Potsdam, the estimated slope is $\widehat{\beta}_1 = 0.05 \pm 0.01$ d/yr when the 90th percentile treshold is employed (Figure 5.12b), and it is $\widehat{\beta}_1 = 0.016 \pm 0.006$ d/yr when the 95th percentile is employed (Figure 5.12d). It makes sense that for the stricter criterion (95th percentile), the WSDI and the slope values are smaller than for the weaker 90th percentile. The upward direction is a robust finding, which also agrees visually with the overall increasing behavior of the occurrence rate for Potsdam (Figure 5.10).

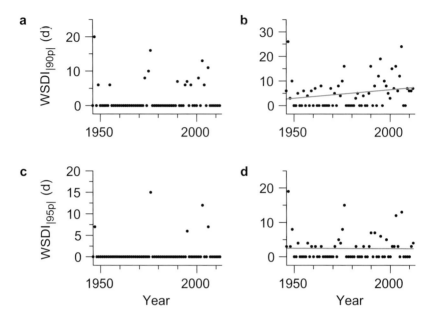

Figure 5.11 Warm spell duration indices for JJA heatwaves at Orléans. The conventional WSDI calculation (a) adopts a duration parameter of six days and an absolute 90th percentile threshold (Table 5.1). Also shown are WSDI time series obtained with (b) duration parameter three days and absolute 90th percentile, (c) duration parameter six days and absolute 95th percentile, and (d) duration parameter three days and absolute 95th percentile. The gray lines (b, d) are linear OLS fits (Section D.1.2).

For Orléans, the estimated slope is $\widehat{\beta}_1 = 0.07 \pm 0.04$ d/yr when the 90th percentile is employed (Figure 5.11b), which is significant. However, the slope is insignificant ($\widehat{\beta}_1 = -0.002 \pm 0.025$ d/yr) for the 95th percentile (Figure 5.11d). The Gaussian assumption for the noise component in the regression equation (Section D.1.2) may be violated because of the still considerable number of zeroes for WSDI. However, also bootstrap techniques (Mudelsee 2014), which "robustify" the regression analysis against such distributional assumptions, yield the same regression results. The absence of trend in WSDI under the stricter 95th percentile criterion may therefore be a real feature. It also has a kind of a visual counterpart in the occurrence rate of JJA heatwaves at Orléans (Figure 5.9b), which decreases strongly during the 1950s and then increases only moderately, but over a long period.

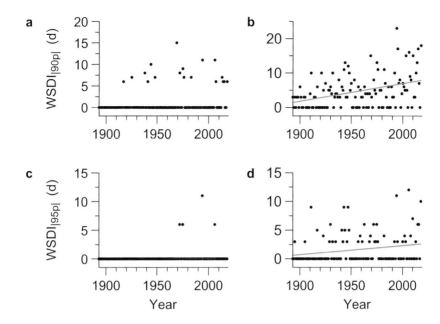

Figure 5.12 Warm spell duration indices for JJA heatwaves at Potsdam. The WSDI series are obtained with (a) duration parameter six days and absolute 90th percentile, (b) duration parameter three days and absolute 90th percentile, (c) duration parameter six days and absolute 95th percentile, and (d) duration parameter three days and absolute 95th percentile. The gray lines (b, d) are linear OLS fits.

However, the WSDI is the WSDI, and the action measure – on the basis of which the occurrence rates are estimated – is the action measure. On the one hand, the two measures react similarly, for example, they both increase when heatwave events get longer. The WSDI slope estimates (90th percentile and three days minimum duration) for Orléans and Potsdam indicate that the increases are in the order of seven days per century and five days per century, respectively. This certainly helps to better understand the changes. On the other hand, the WSDI misses the impact aspect, that means, the size of the temperature exceedance. This is where the action measure has an advantage. Another point – crucial for the risk analyst – is that the action measure delivers POT data in the form of event times and magnitudes. This allows to estimate time-dependent occurrence rates and thus to quantify the risk.

The warm days index is another measure for the analysis of the same datasets from Orléans and Potsdam. We play with the upper threshold

90 Heatwaves and Cold Spells

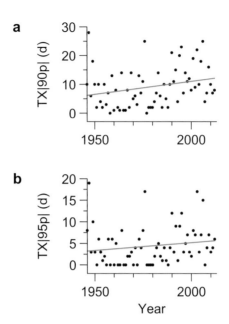

Figure 5.13 Warm days indices for JJA heatwaves at Orléans. The series are obtained with (a) absolute 90th percentile and (b) absolute 95th percentile. The gray lines are linear OLS fits.

parameter (90th and 95th percentiles) to achieve robustness. There is no minimum duration parameter to play with. Figures 5.13 and 5.14 show the results.

Straight lines fitted to the TX|90p| and TX|95p| series reveal significantly increasing trends. For Orléans, the estimated slope is $\widehat{\beta}_1 = 0.09 \pm 0.04$ d/yr when the 90th percentile treshold is employed (Figure 5.13a), and it is $\widehat{\beta}_1 = 0.04 \pm 0.03$ d/yr when the 95th percentile is employed (Figure 5.13b). For Potsdam and the 90th percentile, the estimated slope is $\widehat{\beta}_1 = 0.06 \pm 0.01$ d/yr (Figure 5.14a), and for the 95th percentile, it is $\widehat{\beta}_1 = 0.036 \pm 0.009$ d/yr (Figure 5.14b).

These uniformly upward trends are a robust finding, confirmed by means of bootstrap techniques. Also the comparison of the values makes sense. TX|90p| employs a weaker heatwave criterion than TX|95p|. Hence, TX|90p| and related slope estimates are larger than when the stricter criterion is used. The Potsdam series has more data points than that from Orléans and yields therefore more accurate estimates.

Note that the warm days index is identical to WSDI if for the calculation of the latter the minimum duration condition is relaxed to one day. Therefore, the

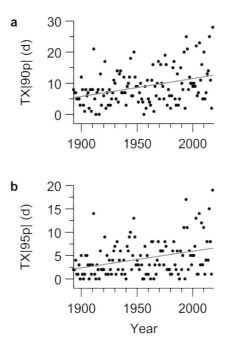

Figure 5.14 Warm days indices for JJA heatwaves at Potsdam. The series are obtained with (a) absolute 90th percentile and (b) absolute 95th percentile. The gray lines are linear OLS fits.

results for warm days are similar to the results for WSDI (minimum duration three days): upward trends. For example, TX|95p| for Potsdam (Figure 5.14b) employs a weaker heatwave criterion (one day) than its WSDI counterpart (three days) (Figure 5.12d). As a result, TX|95p| is larger, and it also yields a larger slope.

5.2.4 Absolute versus Calendar Day Percentiles

The index TX|90p|, calculated with the absolute 90th percentile, is not the same as TX90p, calculated with the calendar day 90th percentile and a five-day window (Section 5.1). The TX90p index, which is conventionally more often used, has a deficit as heatwave measure because it misses aspects of the absolute values. However, indices calculated with calendar day percentiles can illuminate other, "relative" hetwave aspects and should therefore not be ignored. Furthermore, the agreement of results obtained from absolute percentiles with those from relative percentiles can serve as a robust indicator of the overall findings, less sensitive to the selection of methodical details.

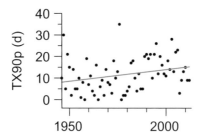

Figure 5.15 Warm days index via calendar day percentiles for JJA heatwaves at Orléans. The data (filled symbols) are from www.ecad.eu (12 February 2019); see also Klein Tank et al. (2002). The gray line is a linear OLS fit.

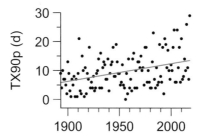

Figure 5.16 Warm days index via calendar day percentiles for JJA heatwaves at Potsdam. The data (filled symbols) are from www.ecad.eu (12 February 2019); see also Klein Tank et al. (2002). The gray line is a linear OLS fit.

Indeed, the general results for TX90p agree well with those for TX|90p| in case of the two European stations. The upward trend in TX|90p| for Orléans (Figure 5.13a) has an estimated slope of $\widehat{\beta}_1 = 0.09 \pm 0.04$ d/yr, which is mirrored by the upward trend in TX90p (Figure 5.15) with $\widehat{\beta}_1 = 0.11 \pm 0.05$ d/yr. (Also the TX90p trends are confirmed by means of bootstrap resampling.) Both estimates are statistically indistinguishable and inform the analyst that, roughly speaking, the heatwave days increase by about ten days per century. Similarly for Potsdam, where the TX|90p| index (Figure 5.14a) has $\widehat{\beta}_1 = 0.06 \pm 0.01$ d/yr and the TX90p index (Figure 5.16) yields the identical result, $\widehat{\beta}_1 = 0.06 \pm 0.01$ d/yr. Six days longer heatwaves at Potsdam after a century.

In our case study of European summer heatwaves, the calendar day percentile threshold (TX90p) gives similar results as the absolute percentile threshold (TX|90p|), even numerically. Evidently, further European temperature measurement stations can be included and the agreement (absolute versus calendar percentiles) mapped. Bear in mind that you learn also from disagreements.

5.3 Case Study: Cold Spells at Potsdam

Also cold extremes can produce damages in natural and cultural systems. Below-freezing temperatures pose a threat to the sensitive parts of a plant (e.g., buds, flowers, and new leaves) during the growing season. One example of a costly event is the April 2007 freeze across the eastern United States, which caused not-inflation-adjusted overall losses of 112 million USD (Warmund et al. 2008).

The Fifth Assessment Report of the IPCC (Hartmann et al. 2013: p. 212 therein) concluded that due to global warming "the number of cold spells has reduced significantly since the 1950s." Indeed, the cited studies (Donat et al. 2013a,b) report significant trends in various indices for cold extremes (upward for TNn and TXn; downward for TN10p, TX10p, and CSDI) in global observation datasets and climate model output. A later study by the same team (Donat et al. 2016) extended the analysis period back in time to 1901 and found similar results (but more variations in the earlier analysis period).

The various indices for cold extremes during the December–January–February (DJF) season at Potsdam (Figure 5.17) roughly agree with the trends reported for the global datasets. The indices are calculated via minimum daily temperature (Table 5.2). Note that December refers to the preceding year, which is combined with January and February to form the boreal winter season. Therefore, the Potsdam series start in 1894, not 1893.

The downward trend in TN10p for station Potsdam (Figure 5.17a) has an estimated slope (confirmed by bootstrap resampling) of $\widehat{\beta}_1 = -0.03 \pm 0.02$ d/yr. CSDI (Figure 5.17b) also has a negative slope, which, however, is not significant: $\widehat{\beta}_1 = -0.008 \pm 0.016$ d/yr. Perhaps the minimum duration parameter, when set to three instead of six days, would yield a significant trend. From the three indices, FD (Figure 5.17c) has the strongest significant negative slope: $\widehat{\beta}_1 = -0.05 \pm 0.03$ d/yr. For the calculation of FD, it seems that the combination of 0 °C as lower threshold for minimum daily temperature and one day minimum duration constitutes a relaxed condition, which leads to numerically high FD values and, compared with the other indices, a better chance to obtain significant results. The values of the significant slope estimates for Potsdam appear somewhat smaller than what is found for many

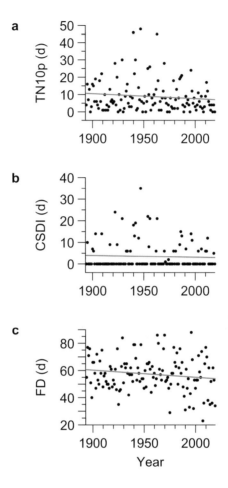

Figure 5.17 Cold extremes during the DJF season at Potsdam. (a) The cold nights index employs the 10th calendar day percentile for the minimum daily temperature. (b) The cold spell duration index employs the 10th calendar day percentile and a minimum duration of six days. (c) The frost day index adopts the freezing point as lower threshold. All data (filled symbols) are from www.ecad.eu (10 April 2019); see also Klein Tank et al. (2002). The series start in 1894 and end in 2019. The gray lines (a–c) are linear OLS fits

regions from the global dataset (Donat et al. 2016), but they are in the same order (a few days less of cold extremes per century).

5.3.1 Late Frost

However, the indices for cold extremes considered in the last paragraph (Figure 5.17) may not be helpful in certain situations. Let us contemplate about the roses I planted last November into the new bed of my office house. Their winter blanket of hilled-up soil was removed during the warm days around the beginning of spring. Three weeks later now (date of writing, 12 April 2019), I fear for the emerged buds because we have freezing nights. This means, the threat is rather the lateness of the frost with regard to the growth start. This is the case not only for my roses but globally for many plant types in the boreal and temperate zones: trees and shrubs in Europe (Bigler and Bugmann 2018; Ma et al. 2018), crops in the United States (Warmund et al. 2008), grapevines in Switzerland (Meier et al. 2018), and so forth. The stunning hypothesis, published over three decades ago (Cannell and Smith 1986), is that with climatic warming – late frost risk increases!

We need a measure for late frost. It has to take the frost date and the start of the growth into account. Growing starts when the mean daily temperature (T_{mean}) over a certain duration is high enough. Donat et al. (2013a), and many others, define what we call the Start of Growing Season (SGS) as the first occurrence of six consecutive days with $T_{mean} > 5$ °C. The Last Frost Day (LFD) in a growing season is obviously defined via the minimum temperature, $T_{min} < 0$ °C. A late frost is the severer the more days have passed since growth start. Hence, we define the Late Frost Severity Index (LFSI) via the equation

$$\text{LFSI} = \text{LFD} - \text{SGS}. \tag{5.1}$$

If SGS > LFD, then there is no late frost event and LFSI is not defined.

Figure 5.18 shows late frost events and the calculation of the LFSI for cold extremes at Potsdam. For the year 2015, T_{mean} was above 5 °C on 15 to 20 March; therefore, SGS = 79 d. If also on 21 March T_{mean} were above 5 °C (which it was not), the SGS value would not change. 7 April 2015 was the last freezing day ($T_{min} = -1.2$ °C). Therefore, LFD = 97 d and LFSI = 18 d.

Note that the calculation of the LFD has to impose a condition on the allowed days in a year in order to exclude frost events that belong already to the next winter season (e.g., in December). For Potsdam (Figure 5.18c), we require that the LFD in a growing season – if it occurs – has to happen until the summer peak (day 183). Analogously, the calculation of the SGS for Potsdam imposes the condition that SGS > 31 d in order to exclude too early (January) warm events. An additional motivation for this condition is that another factor for growth – sunlight – should be favorable. It is clear that other geographical regions (e.g., in the Southern Hemisphere) may require other settings than Potsdam.

To summarize our assessment of the LFSI, its advantages are that it takes magnitude and duration into account (Requirement 1) and needs only

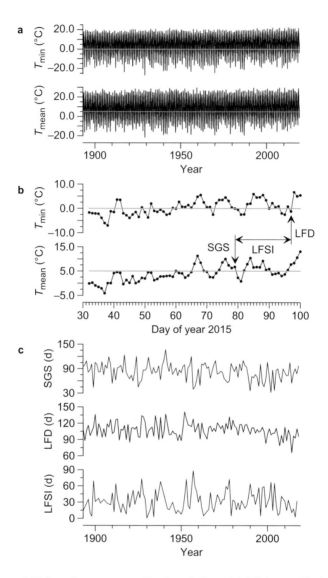

Figure 5.18 Late frost events at Potsdam: indices. (a) Minimum (T_{min}) and mean (T_{mean}) daily surface-air temperature at Potsdam during the interval from 1 January 1893 to 31 December 2018. The minimum is taken from the three daily temperature readings at 7:00 a.m., 2:00 p.m., and 9:00 p.m.; the mean is calculated from these readings, whereby the 9:00 p.m. value is used twice. Data from Friedrich-Wilhelm Gerstengarbe (Potsdam Institute for Climate Impact Research, Germany, personal communication, 2014) and www.pik-potsdam.de/services/climate-weather-potsdam/climate-diagrams (6 February 2019). Also shown (gray horizontal lines) are the thresholds for T_{min} (0 °C) and T_{mean} (5 °C). (b) Zoom on interval from day 32 (1 February) to day 100 (10 April) of the year 2015, illustrating the definitions of the SGS, LFD, and LFSI indices. (c) Full series of these indices

temperature data for its calculation (Requirement 3). One critical point is that the application of the LFSI to widely distributed geographical locations may not be straightforward, which would mean a violation of Requirement 2. Another problem is that the calculation of the LFSI has to employ several parameters (minimum duration, thresholds, and allowed days in a year), which brings conflicts with Requirement 4. One option to improve the situation is to take direct observations of the SGS in the field (as I did with my roses). This may yield more robust LFSI estimates. However, these SGS observations are expensive and have likely to be taken separately for the different plant types of interest. Another option is to utilize more advanced empirical SGS models (Cannell and Smith 1986), which then, however, may have new data demands and bring conflicts with Requirement 3.

The LFSI data have another, risk-analytical advantage. They quantify the size of an (adverse) event and give also its date (via the LFD). That means, the LFSI has the POT data format (Section 2.2). This allows to test the hypothesis that with climatic warming the late frost risk increases.

Figure 5.19 summarizes our attempts to test the late frost hypothesis by means of temperature series from Potsdam. Bandwidth selection ($h = 12$ a) is motivated by the finding that smaller bandwidths yield more – but statistically insignificant – wiggles of the occurrence rate. From the 126 years of observation, 102 do show late frost events according to Eq. (5.1). If we put all events into one magnitude class ($m = 102$), then the one-sided Cox–Lewis test (Section 3.2.5) does not allow to reject the null hypothesis "no trend" at any decent level of significance ($P = 0.26$).

We prefer a selection of two magnitude classes for late frosts at Potsdam. The strong events (LFSI > 14 d, $m = 83$), which are also by far the costlier events, do indeed reveal a long-term increasing trend (Figure 5.19b). However, there are some caveats. First, the trend seems to switch from upward to downward at around 1985. Why? Second, the P-value for the Cox–Lewis test is still rather large ($P = 0.18$). Third, the weak events (LFSI ≤ 14 d, $m = 19$) seem to indicate a contrasting trend behavior (downward–upward) (Figure 5.19a). Also the focus on even stronger events (LFSI > 21 d, $m = 65$) does not produce definitive answers (results not shown). The major conclusion is that although the Potsdam temperature time series have a superior, homogeneous data quality, the nonstationary risk analyses still are affected by large uncertainties (i.e., wide confidence bands and large P-values). This prevents a closer inspection of the hypothesis by Cannell and Smith (1986) that with climatic warming the late frost risk increases. An improvement for the assessment of the hypothesis may come when the LFSI for Potsdam can be calculated with directly observed SGS data.

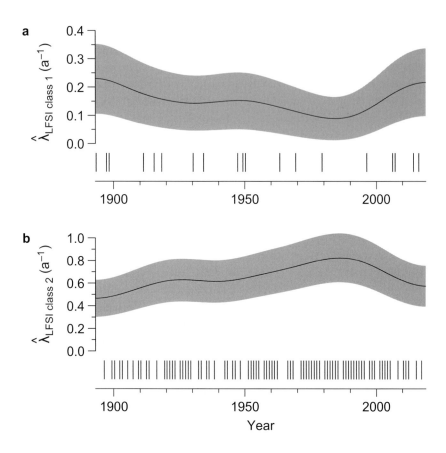

Figure 5.19 Late frost events at Potsdam: occurrence rates. Shown are two magnitude classes (a, class 1, LFSI $\leq$ 14 d; b, class 2, LFSI > 14 d). The number of events is (a) $m = 19$ and (b) $m = 83$. The occurrence rates (solid lines) with 90% confidence bands (shaded) are estimated by means of a Gaussian kernel with a bandwidth of $h = 12$ a and 2000 bootstrap resamples.

5.4 Outlook

We have applied in this chapter on heatwaves and cold spells the statistical methods (Chapter 3) to daily temperature time series from two stations in Europe, Orléans in France and Potsdam in Germany. Let us assess how these case studies can be extended in the two research directions space and time. The third direction, time resolution, is already adequate: we have daily values, and it is difficult to study hourly extreme anomalies against the strong daily cycle.

First, space. The mantra "More stations have to be analyzed to achieve a comprehensive geographical overview" applies also to our approach to understand heatwaves and cold spells in Europe. To appease ourselves, we note that the distance between Orléans and Potsdam is nearly thousand kilometers, which should make the information contents of both stations statistically independent of each other. For example, Orléans in the west should be stronger influenced by the North Atlantic Oscillation (NAO) than Potsdam in the east. The NAO, indexed by the air-pressure difference between the North Atlantic's north (e.g., Iceland) and its south (e.g., Lisbon), describes the strength and direction of the westerlies and their influence on the weather in Europe (Hurrell 1995). Despite their spatial distance, Orléans and Potsdam report comparable upward trends in heatwave occurrence over the past decades (Figures 5.9 and 5.10). That means, we deal with a phenomenon of large spatial extent. On the other hand, there remain regional differences. The August 2003 heatwave set a record for Orléans, but not for Potsdam (Section 5.2).

Second, time. The extension of the Potsdam series (since 1893) back to the seventeenth century (invention of thermometer) may in principle be possible. For example, there exist thermometer readings from central England, which start in 1659 (Manley 1974), or there is a report about the heat at Plymouth in July 1757 (Huxham 1758). Practically, however, it seems nearly impossible to beat Potsdam in terms of time resolution (daily), completeness, and homogeneity (Chapter 2). The quality level of Potsdam is unachieved also by documentary or natural datasets. Documentary data can certainly be employed in a quantitative manner. For example, the European heatwave in summer 1540 has been found to have had a magnitude comparable to the event in summer 2003 (Wetter et al. 2014; Orth et al. 2016). However, usage of documentary data means also a risk of perception bias and underreporting of minor events (Chapter 2). Also proxy temperature data from natural archives (Appendix B) cannot offer daily resolution and completeness over a long time interval. And also here, minor temperature extremes may be difficult to detected against the proxy and measurement noise (Appendix A). Climatologists have to live with the fact that the price to pay for the extension of the heatwave and cold spell records back in time is a lower data quality: lower resolution, reduced completeness, and the risk of missed minor events. Still, the paleo-temperature extremes can be studied using advanced methods (Chapter 3), but the data do not have to be transformed into the sophisticated indices (Tables 5.1 and 5.2).

The indices. As said, they are not required to study paleo time series, but in high-quality data they can illuminate aspects of relevance for the practitioner. We think that the new impact action measures, which take the magnitude and duration aspects of a heatwave simultaneously into account (Table 5.1),

are helpful for health professionals and others. We further think that the
new indicator (LFSI) for late frost severity (Table 5.2) has the potential to
help farmers and others. The AMICE software (Section F.2) performs the
calculation of these (and other) indices. The performance of the new and the
old indices should be compared in Monte Carlo experiments with artificial
data (Section D.1.4) to see what can be done, and what not. The future task
for climate researchers is then to (1) take high-quality measurement data or
climate model output, (2) calculate the suite of index series, (3) apply trend
and risk estimation procedures, and (4) represent and summarize the results
(which may include many multi-panel figures) in an accessible form to their
peers and the users of their research output. One example paper is that by Donat
et al. (2016).

5.5 Summary for the Risk Analyst

Economically relevant are the current situation and the near future (the next
five years). For strategic decisions, also the next few decades count. The
ideal resolution of temperature input data (maximum, minimum, and mean)
for this time range is daily or better (i.e., smaller). Note that the input data
are obtained from several temperature readings per day. Temperature shows
smaller spatial variations than precipitation (Chapter 1). Still, a certain number
of geographical sites has to be studied to achieve a comprehensive analysis of
a larger region.

The risk of warm extremes is rather well understood, at least better than
flood risk (Chapter 4) or storm risk (Chapter 6). The reason is that temperature
is more directly related to GHG concentrations than precipitation or wind
speed (Chapter 1). The concern that the rising GHG concentrations lead to
more and stronger warm extremes is well justified. The concern is confirmed
in the case study for the European weather stations Orléans in France and
Potsdam in Germany.

However, a data-analytical challenge is posed by the fact that a heatwave
(warm extreme) bears also the duration aspect. Longer heatwaves cost more.
To meet the challenge, climatologists have devised heatwave indices, which
are calculated from the temperature and the time values (days). This textbook
presents also a new index, called action measure, which seems to successfully
combine the magnitude and duration aspects (Table 5.1). The action measure
reveals that the heatwave from 2 to 14 August 2003 was the strongest event at
Orléans since the start of the regular observations in 1946.

The action measure has another, risk-analytical advantage. It quantifies the
magnitude of a threshold exceedance event (via the product of duration and

exceedance) and it also gives the event date (via the center of the exceedance interval). That means, the action measure has the POT data format (Section 2.2). This allows to examine heatwave events in a nonstationary framework and estimate trends in occurrence (Section 3.2). This method has allowed us to quantify for the European heatwaves the upward trends over the recent decades and, on basis of the long series from Potsdam, to detect a high in occurrence during the 1940s (Section 5.2.2).

The risk-analytical methodology has to recognize that there are two parameters to be appropriately adjusted. First, the level of the temperature threshold and second, the minimum duration of a heatwave event. From the case studies we learn that the 90th absolute percentile as threshold and a minimum duration of three days can be a good choice. However, it is mandatory to play with the parameter setting and study the sensitivity of results in order to obtain a robust assessment. This applies not only to the action measure but to all indices (warm or cold extremes) listed in Tables 5.1 and 5.2. A selection of the 95th percentile or a minimum duration of six days may be an interesting option if the data size is high enough (i.e., at least several thousand).

The methodology has also to listen to the needs of the end users and fit the setting of parameters and the design of the indices accordingly. It appears in this regard more appropriate to consider absolute thresholds than calendar-day anomalies. For example, the 1 June 1981 at Orléans with +5.61 °C above what is normal has an absolute value of 26.6 °C, which is not very relevant for human health.

Cold spells are equally well understood as heatwaves. Rising GHG concentrations are expected to lead to fewer cold extremes. This is confirmed in the case study for Potsdam (Section 5.3). However, there is the hypothesis that the risk of late frost events increases with climatic warming (Cannell and Smith 1986). The growing season of plants starts earlier if the mean daily temperature rises. The risk that after growth start the minimum daily temperature drops below zero is still there. Hence, late frosts (after the growth start) may become more frequent.

A suitable index is required to measure the severity of late frosts. This textbook presents a new index (LFSI), which is defined (Table 5.2) as the time span between growth start and the last freezing day in a season. The longer the span, the severer the impact of the frost.

Also the LFSI series have the POT data format and can be examined in a nonstationary risk analysis framework. Although the Potsdam temperature time series have a superior data quality and go back as far as 1893 – the risk analysis results are still affected by large uncertainties. This prevents a definitive test of the late frost hypothesis. The situation can be improved by the study of directly observed growth start data. These data should depend on the type of crop, and the existing time series likely do not go back as far as 1893. However, this is not a problem since it is the recent period that is relevant for the risk analyst. We expect progress on the late frost hypothesis for Potsdam and elsewhere from the analysis of more data (from agriculture) by means of the LFSI.

The stationarity assumption for temperature is obsolete in this world under climate change. Under nonstationarity, the task of predicting the time-dependent occurrence rate of heatwaves or cold spells into the near future can be performed via two approaches. First, by means of extrapolation of the curve, which has the advantage to "let the data speak for themselves." Second, note that temperature changes and the driving factors are well understood, at least better than changes in precipitation or wind speed. This should make risk prediction of heatwaves and cold spells by means of climate models (Section C.1) feasible.

Finally, a hint on analytical methodology. The heatwave and cold spell indices (Tables 5.1 and 5.2) are designed to capture the relevant aspects of extremes from daily time series. Similar indices exist for precipitation extremes (see the literature in the Reading Material).

Reading Material

The Fifth IPCC Report (Hartmann et al. 2013) and more recent papers, such as that by Donat et al. (2016), include assessments of extremes in temperature and precipitation. Zhang et al. (2011) present indices for monitoring changes in temperature and precipitation extremes, while Perkins and Alexander (2013) focus on heatwave indices. Late frost risk and its association with climatic warming have been raised long ago (Cannell and Smith 1986) and are topics actively researched today (Bigler and Bugmann 2018).

The causal attribution of heatwave occurrence and record temperatures to climatic warming trends can be achieved with the help of climate models (Otto et al. 2012; National Academies of Sciences, Engineering, and Medicine 2016) or in a statistical framework (Rahmstorf and Coumou 2011). These three publications study also the 2010 Russian heatwave. Mudelsee (2019) employs the instrumental time series of global surface temperature for the illustration of trend estimation methods. Winter and Tawn (2016) present a statistical model for the joint aspects of duration and temperature threshold, which they apply to the summer 2003 heatwave at Orléans.

Popper (1959) is a classic book, which introduces falsifiability as criterion for science.

6

Hurricanes and Other Storms

A cyclone is an atmospheric low-pressure system with a closed, roughly circular wind motion that is clockwise in the Southern Hemisphere and counterclockwise in the Northern (Neuendorf et al. 2005). A hurricane is a tropical cyclone in the North Atlantic–West Indies region with near-surface wind speed equal to or larger than 64 kn (Elsner and Kara 1999). Expressed in other units, this threshold value equals 12 Beaufort or about 33 $\mathrm{m\,s}^{-1}$. Hurricane Katrina in August 2005, which made landfall and hit the United States, was an extraordinarily powerful and deadly hurricane with at least 1833 human victims and not-inflation-adjusted overall losses of clearly over 100 billion USD (Knabb et al. 2011). Also other storm types than hurricanes may be deadly and costly events. It is clearly the magnitude (maximum wind speed) that determines the impacts of a storm. The duration aspect is less relevant.

The case studies focus on the occurrence rate. Does the risk change over time, are there trends in the occurrence of hurricanes or other storms of a certain magnitude? The first study is on observed hurricanes making landfall in the United States between 1851 and 2018 (Section 6.1). The second analysis extends the view on hurricanes near Boston back in time by nearly thousand years (Section 6.2). The third study is on storms at Potsdam between 1893 and 2018 (Section 6.3). Finally, we indicate current research directions (Section 6.4) and give a short summary for risk analysts (Section 6.5).

6.1 Case Study: US Landfalling Hurricanes

Second to hurricane Katrina in August 2005 as costliest event, according to the National Hurricane Center of the United States (see Reading Material), ranks hurricane Harvey in August 2017. The database (Section 6.1.1) comprises US landfalling hurricanes since 1851. Landfall means better chances to get reported. Therefore, data inhomogeneities due to missed events should be smaller than if we considered all hurricanes, that is, including those that stay and decay on the ocean (where they originate). Landfall also means higher costs and increased relevance for the insurance industry. The risk analysis (Section 6.1.2) compares the long-term observational record with the recent decades (when the costliest events happened). The analysis includes a sensitivity study to assess the influence of the data quality on the result.

6.1.1 Database

The hurricane season is from 1 June to 30 November. This interval encompasses the vast majority of tropical wind activity. No hurricane is away from the list (Table 6.1) for the bureaucratic reason that it happened outside of the official season.

The hurricane events are conventionally classified according to the Saffir–Simpson hurricane wind scale. The relevant meteorological variable is the highest average wind speed, v, that is sustained at the surface (10 m) over a one-minute time span. The largest event (category 5) is defined by $v \geq 137$ kn. One knot (kn) equals $463/900 \approx 0.514$ m s^{-1}. The other wind-speed boundaries are: category 4, $v \geq 113$ kn; category 3, $v \geq 96$ kn; category 2, $v \geq 83$ kn; and category 1, $v \geq 64$ kn. The values reported for a hurricane (Table 6.1) are conventionally rounded to multiples of 5 kn (data since 1886) or 10 kn (data before).

It is "easier" for a hurricane to sustain a high wind speed for one minute than for ten minutes. For example, the storms at Potsdam in Germany (Section 6.3) are defined via a ten-minute time span, and the maximum observed there during the full interval (1893–2018) was 20.8 m s^{-1}. Such a value – would not qualify even as a category-1 hurricane. This shows that the comparability of absolute wind-speed values from different sustained measurement spans is limited.

One motivation to focus on landfalling hurricanes is data homogeneity. Still, the lack of continuously populated coastal regions in the earlier part of the observation interval means that there is a risk of missed hurricane events, especially of lower category, for the period from 1851 to 1910 (Landsea et al. 2004). With the technological inventions and achievements of the twentieth century (radio, airplane, weather satellite, and measuring instruments), the data situation improved. Also new documentary data about hurricanes are discovered, analyzed, and incorporated into the database.

Table 6.1. *US landfalling hurricanes, 1851–2018. The listed date refers to the peak in the wind speed of a hurricane. The database is from the National Oceanic and Atmospheric Administration, Atlantic Oceanographic and Meteorological Laboratory, Hurricane Research Division. The data for 1851–2017 were downloaded from www.aoml.noaa.gov/hrd/hurdat/ Data_Storm.html (6 June 2019), the data for 2018 from www.aoml.noaa.gov/ keynotes/PDF-Files/Nov-Dec2018.pdf (4 June 2019). Data corrections for 1954 1963 (Delgado et al. 2018) are taken into account. On 9 September 1971, hurricane Fern peaked at 6:00 a.m. and Edith at 6:00 p.m. The time series of events is shown in Figure 6.1.*

Year	Month	Day	Category	Maximum wind speed (kn)	Name
1851	Jun.	25	1	80	
1851	Aug.	23	3	100	Great Middle Florida
1852	Aug.	26	3	100	Great Mobile
1852	Sep.	12	1	70	
1852	Oct.	9	2	90	Middle Florida
1853	Oct.	21	1	70	
1854	Jun.	26	1	70	
1854	Sep.	8	3	100	Great Carolina
1854	Sep.	18	2	90	Matagorda
1855	Sep.	16	3	110	Middle Gulf Shore
1856	Aug.	10	4	130	Last Island
1856	Aug.	31	2	90	Southeastern States
1857	Sep.	13	2	90	
1858	Sep.	16	1	80	New England
1859	Sep.	16	1	70	
1859	Oct.	28	1	80	
1860	Aug.	11	3	110	
1860	Sep.	15	2	90	
1860	Oct.	2	2	90	
1861	Aug.	16	1	70	Key West
1861	Sep.	27	1	70	Equinoctial
1861	Nov.	2	1	70	Expedition
1865	Sep.	13	2	90	Sabine River
1865	Oct.	23	2	90	
1866	Jul.	15	2	90	
1867	Jun.	22	1	70	
1867	Oct.	4	2	90	Galveston
1869	Aug.	17	2	90	Lower Texas Coast
1869	Sep.	5	1	70	
1869	Sep.	8	3	100	Eastern New England
1869	Oct.	4	2	90	Saxby's Gale
1870	Jul.	30	1	70	Mobile
1870	Oct.	10	1	70	Twin Key West (I)

Table 6.1. *(Cont.)*

Year	Month	Day	Category	Maximum wind speed (kn)	Name
1870	Oct.	20	1	80	Twin Key West (II)
1871	Aug.	17	3	100	
1871	Aug.	25	2	90	
1871	Sep.	6	1	70	
1873	Sep.	19	1	70	
1873	Oct.	7	3	100	
1874	Sep.	28	1	80	
1875	Sep.	16	3	100	
1876	Sep.	17	1	80	
1876	Oct.	20	2	90	
1877	Sep.	18	1	70	
1877	Oct.	3	3	100	
1878	Sep.	10	2	90	
1878	Oct.	23	2	90	
1879	Aug.	18	3	100	
1879	Aug.	23	2	90	
1879	Sep.	1	3	110	
1880	Aug.	13	3	110	
1880	Aug.	29	2	90	
1880	Sep.	9	1	70	
1880	Oct.	8	1	70	
1881	Aug.	28	2	90	
1881	Sep.	9	2	90	
1882	Sep.	10	3	110	
1882	Oct.	11	1	70	
1883	Sep.	11	2	90	
1885	Aug.	25	2	90	
1886	Jun.	14	2	85	
1886	Jun.	21	2	85	
1886	Jun.	30	2	85	
1886	Jul.	19	1	70	
1886	Aug.	20	4	130	Indianola
1886	Sep.	23	1	80	
1886	Oct.	12	3	105	
1887	Jul.	27	1	75	
1887	Aug.	20	1	65	
1887	Sep.	21	1	75	
1887	Oct.	19	1	75	
1888	Jun.	17	1	70	
1888	Aug.	16	3	110	
1888	Oct.	11	2	95	
1889	Sep.	23	1	70	
1891	Jul.	5	1	80	
1891	Aug.	24	1	70	
1893	Aug.	24	1	75	Midnight Storm

Table 6.1. *(Cont.)*

Year	Month	Day	Category	Maximum wind speed (kn)	Name
1893	Aug.	28	3	100	Sea Islands
1893	Sep.	7	2	85	
1893	Oct.	2	4	115	Chenier Caminanda
1893	Oct.	13	3	105	
1894	Sep.	25	2	90	
1894	Oct.	9	3	105	
1895	Aug.	30	1	65	
1896	Jul.	7	2	85	
1896	Sep.	10	1	70	
1896	Sep.	29	3	110	
1897	Sep.	13	1	75	
1898	Aug.	2	1	70	
1898	Aug.	31	1	75	
1898	Oct.	2	4	115	
1899	Aug.	1	2	85	
1899	Aug.	18	3	105	
1899	Oct.	31	2	95	
1900	Sep.	9	4	120	Galveston
1901	Jul.	11	1	70	
1901	Aug.	14	1	75	
1903	Sep.	13	1	80	
1903	Sep.	16	1	70	
1904	Sep.	14	1	70	
1904	Oct.	17	1	70	
1906	Jun.	17	1	75	
1906	Sep.	17	1	80	
1906	Sep.	27	2	95	
1906	Oct.	18	3	105	
1908	Jul.	31	1	70	
1909	Jun.	29	2	85	
1909	Jul.	21	3	100	Velasco
1909	Aug.	27	1	65	
1909	Sep.	21	3	105	Grand Isle
1909	Oct.	11	3	100	
1910	Sep.	14	2	90	
1910	Oct.	18	2	95	
1911	Aug.	11	1	70	
1911	Aug.	28	2	85	
1912	Sep.	14	1	65	
1912	Oct.	16	2	85	
1913	Jun.	28	1	65	
1913	Sep.	3	1	75	
1913	Oct.	8	1	65	
1915	Aug.	1	1	65	
1915	Aug.	17	4	115	Galveston

Table 6.1. *(Cont.)*

Year	Month	Day	Category	Maximum wind speed (kn)	Name
1915	Sep.	4	1	80	
1915	Sep.	29	3	110	New Orleans
1916	Jul.	5	3	105	
1916	Jul.	14	2	95	
1916	Aug.	18	4	115	
1916	Oct.	18	2	95	
1917	Sep.	29	3	100	
1918	Aug.	6	3	105	
1918	Aug.	24	1	65	
1919	Sep.	10	4	130	
1920	Sep.	22	2	85	
1921	Jun.	22	1	80	
1921	Oct.	25	3	100	Tampa Bay
1923	Oct.	16	1	70	
1924	Aug.	26	1	65	
1924	Sep.	15	1	75	
1924	Oct.	21	1	80	
1926	Jul.	28	2	90	
1926	Aug.	25	3	100	
1926	Sep.	18	4	125	Great Miami
1926	Oct.	21	1	75	
1928	Aug.	8	2	85	
1928	Sep.	17	4	125	Lake Okeechobee
1929	Jun.	28	1	80	
1929	Sep.	28	3	100	
1932	Aug.	14	4	130	Freeport
1932	Sep.	1	1	75	
1933	Aug.	5	1	80	
1933	Aug.	23	1	80	
1933	Sep.	4	3	110	
1933	Sep.	5	3	110	
1933	Sep.	16	2	85	
1934	Jun.	16	2	85	
1934	Jul.	25	1	75	
1934	Sep.	8	1	65	
1935	Sep.	3	5	160	Labor Day
1935	Nov.	4	2	85	
1936	Jun.	27	1	70	
1936	Jul.	31	2	90	
1936	Sep.	18	1	75	
1938	Aug.	15	1	65	
1938	Sep.	21	3	105	Great New England
1939	Aug.	13	1	65	
1940	Aug.	7	2	85	
1940	Aug.	11	2	85	

Table 6.1. *(Cont.)*

Year	Month	Day	Category	Maximum wind speed (kn)	Name
1941	Sep.	23	3	110	
1941	Oct.	6	2	85	
1942	Aug.	21	1	65	
1942	Aug.	30	3	100	
1943	Jul.	27	2	90	
1944	Aug.	1	1	70	
1944	Sep.	15	2	90	Great Atlantic
1944	Oct.	18	3	105	
1945	Jun.	24	1	70	
1945	Aug.	27	3	100	
1945	Sep.	15	4	115	
1946	Oct.	8	1	75	
1947	Aug.	24	1	70	
1947	Sep.	16	4	115	
1947	Oct.	15	2	90	
1948	Sep.	4	1	70	
1948	Sep.	22	4	115	
1948	Oct.	5	2	90	
1949	Aug.	26	4	115	
1949	Oct.	4	2	95	
1950	Aug.	31	1	75	Baker
1950	Sep.	5	3	105	Easy
1950	Oct.	18	4	115	King
1952	Aug.	31	2	85	Able
1953	Aug.	14	1	80	Barbara
1953	Sep.	26	1	80	Florence
1953	Oct.	9	1	75	Hazel
1954	Aug.	31	3	100	Carol
1954	Sep.	11	2	95	Edna
1954	Oct.	15	4	115	Hazel
1955	Aug.	12	2	85	Connie
1955	Sep.	19	2	90	Ione
1956	Sep.	24	1	75	Flossy
1957	Jun.	27	3	110	Audrey
1958	Sep.	27	3	110	Helene
1959	Jul.	9	1	65	Cindy
1959	Jul.	25	1	75	Debra
1959	Sep.	29	4	115	Gracie
1960	Sep.	10	4	125	Donna
1960	Sep.	15	1	70	Ethel
1961	Sep.	11	4	125	Carla
1964	Aug.	23	2		Cleo
1964	Sep.	6	2		Dora
1964	Oct.	1	3		Hilda
1964	Oct.	14	2		Isbell

Table 6.1. *(Cont.)*

Year	Month	Day	Category	Maximum wind speed (kn)	Name
1965	Sep.	10	3		Betsy
1966	Jun.	8	2		Alma
1966	Sep.	28	1		Inez
1967	Sep.	20	3		Beulah
1968	Oct.	19	2		Gladys
1969	Aug.	16	5		Camille
1969	Sep.	9	1		Gerda
1970	Aug.	3	3		Celia
1971	Sep.	9	1		Fern
1971	Sep.	9	2		Edith
1971	Sep.	14	1		Ginger
1972	Jun.	19	1		Agnes
1974	Sep.	2	3		Carmen
1975	Sep.	23	3		Eloise
1976	Aug.	9	1		Belle
1977	Sep.	5	1		Babe
1979	Jul.	11	1		Bob
1979	Aug.	30	2		David
1979	Sep.	12	3		Frederic
1980	Aug.	7	3	100	Allen
1983	Aug.	18	3	100	Alicia
1984	Sep.	13	2	95	Diana
1985	Jul.	25	1	65	Bob
1985	Aug.	15	1	80	Danny
1985	Sep.	2	3	100	Elena
1985	Sep.	27	2	90	Gloria
1985	Oct.	29	1	75	Juan
1985	Nov.	21	2	85	Kate
1986	Jun.	26	1	75	Bonnie
1986	Aug.	17	1	65	Charley
1987	Oct.	12	1	65	Floyd
1988	Sep.	9	1	70	Florence
1989	Aug.	1	1	70	Chantal
1989	Sep.	22	4	120	Hugo
1989	Oct.	16	1	75	Jerry
1991	Aug.	19	2	90	Bob
1992	Aug.	24	5	145	Andrew
1993	Aug.	31	3	100	Emily
1995	Aug.	3	2	85	Erin
1995	Oct.	4	3	100	Opal
1996	Jul.	12	2	90	Bertha
1996	Sep.	6	3	100	Fran
1997	Jul.	19	1	70	Danny
1998	Aug.	27	2	95	Bonnie
1998	Sep.	3	1	70	Earl

Table 6.1. *(Cont.)*

Year	Month	Day	Category	Maximum wind speed (kn)	Name
1998	Sep.	25	2	90	Georges
1999	Aug.	23	3	100	Bret
1999	Sep.	16	2	90	Floyd
1999	Oct.	18	2	95	Irene
2002	Oct.	3	1	80	Lili
2003	Jul.	15	1	80	Claudette
2003	Sep.	18	2	90	Isabel
2004	Aug.	3	1	70	Alex
2004	Aug.	13	4	130	Charley
2004	Aug.	29	1	65	Gaston
2004	Sep.	5	2	90	Frances
2004	Sep.	16	3	105	Ivan
2004	Sep.	26	3	105	Jeanne
2005	Jul.	6	1	65	Cindy
2005	Jul.	10	3	105	Dennis
2005	Aug.	29	3	110	Katrina
2005	Sep.	15	1	65	Ophelia
2005	Sep.	24	3	100	Rita
2005	Oct.	24	3	105	Wilma
2007	Sep.	13	1	80	Humberto
2008	Jul.	23	1	75	Dolly
2008	Sep.	1	2	90	Gustav
2008	Sep.	13	2	95	Ike
2011	Aug.	27	1	75	Irene
2012	Aug.	29	1	70	Isaac
2012	Oct.	29	1	65	Sandy
2014	Jul.	4	2	85	Arthur
2016	Sep.	2	1	70	Hermine
2016	Oct.	8	2	85	Matthew
2017	Aug.	26	4	115	Harvey
2017	Sep.	10	4	115	Irma
2017	Oct.	8	1	65	Nate
2018	Sep.	11	4	130	Florence
2018	Oct.	10	4	135	Michael

The aim of the Atlantic Hurricane Database Reanalysis project (Landsea et al. 2004) is to compile a good hurricane database – as full as possible and of good homogeneity. The latest paper from the project (Delgado et al. 2018) shows the reanalysis up to 1963. The absence of reported wind-speed maxima for the interval from 1964 to 1979 (Table 6.1) may indicate that the project has not yet advanced so far and that the data quality for that interval is reduced.

Another challenge with hurricane data is that in principle a hurricane event constitutes a wind-speed extreme not only in time but also in space.

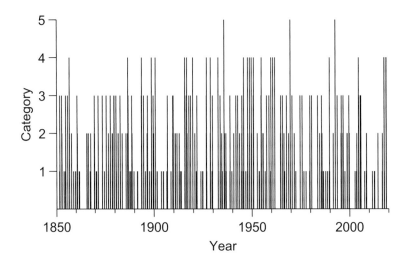

Figure 6.1 US landfalling hurricanes, 1851–2018, time series. The events are listed in Table 6.1. Category 5 has 3 events, category 4 has 26 events, category 3 has 63 events, category 2 has 81 events, and category 1 has 119 events.

A hurricane forms (Emanuel 1987), moves, receives and dissipates its energy, has possibly several ups and downs in sustained wind speed over time, makes landfall or not, and, finally, decays into a weak extra- or intratropical cyclone. To squeeze this time- and space-dependent behavior into a single number (maximum wind speed, v) inevitably introduces uncertainties into the database. Delgado et al. (2018) report uncertainties of v, which are between 15 kn for the earlier and 9 kn for the later part of the observation interval. The usage of the Saffir–Simpson hurricane wind scale with five categories reduces the influence of these uncertainties. This is in analogy to the Elbe flood record with three categories (Section 2.1.1).

The time series for the risk analysis consists of 292 events in total (Figure 6.1). The data are of type "event times" (Section 2.1) and have the hurricane category as magnitude measure. The observation interval is from 1 June 1851 to 30 November 2018.

6.1.2 Risk Analysis

In the nonstationary analysis framework, we estimate the time-dependent hurricane occurrence rate (Figure 6.2). Given the uncertainties of the database (Section 6.1.1), the objective of the analysis is the robustness of inferred trends in the occurrence rate. Therefore we play with the magnitude aspect

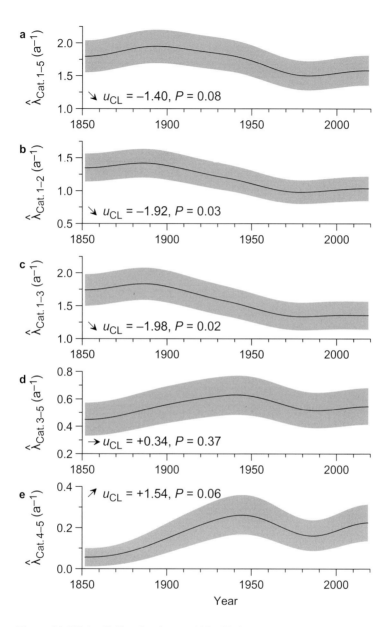

Figure 6.2 US landfalling hurricanes, 1851–2018, occurrence rate estimation. Shown are the results for various category ranges (a, 1–5; b, 1–2; c, 1–3; d, 3–5; e, 4–5). The occurrence rates (solid lines) with 90% confidence bands (shaded) are estimated by means of a Gaussian kernel with a bandwidth of (a–e) $h = 20$ a. The number of resamples is (a–e) 2000. Also given are the results for the Cox–Lewis test statistic, u_{CL}, and the one-sided P-value; arrows indicate significant (90%) upward, downward, or insignificant (horizontal arrow) trends in occurrence rate. Cat., category.

and study various groups of hurricane categories. We employ the Cox–Lewis test (Section 3.2.5) to scrutinize inferred occurrence-rate trends.

Boundary bias correction via the reflection method of pseudodata generation (Section 3.2.3) is relevant for both observation-interval boundaries (1851, 2018) since close to the boundaries certain numbers of hurricane events are found (Table 6.1). The cross-validation functions (not shown) do not exhibit well-defined local minima and are thus not helpful for guiding bandwidth selection. One reason may be existing autocorrelation effects (as in the monsoon study in Section 4.2.2). More helpful for the purpose of the selection of the bandwidth, h, is to play with h and find out by means of the confidence band what the real features of the occurrence rate curve are, and what just insignificant wiggles (as in the late-frost study in Section 5.3.1). The outcome of this exercise is that the selection of a value of $h = 20$ a yields a good solution to the smoothing problem for the hurricane data. The number of resamples ($B = 2000$) follows the usual recommendation (Section 3.2.4).

The full range of hurricane categories, 1–5, yields the largest sample size ($m = 292$). This largest set of events yields a significant long-term downward trend estimate (Figure 6.2a). The occurrence rate at around the start of the observation interval in 1851 is $\widehat{\lambda}(T) \approx 1.8 \text{ a}^{-1}$, and the 90% confidence interval (CI) for this estimate is $[1.5 \text{ a}^{-1}; 2.0 \text{ a}^{-1}]$. The rate at around the end of the observation interval in 2018 is $\widehat{\lambda}(T) \approx 1.6 \text{ a}^{-1}$ $[1.4 \text{ a}^{-1}; 1.8 \text{ a}^{-1}]$. The Cox–Lewis test confirms the significance of the downward trend ($P = 0.08$). There may be hidden a slowdown of the trend and even a turn to an increasing trend during the 1980s (Figure 6.2a), but the confidence band is too wide to allow speculations about this.

The restriction to lower hurricane categories, 1 to 2 with $m = 200$ events or 1 to 3 with $m = 263$ events, also leads to significant downward trends (Figure 6.2b, c). The downward character is even more pronounced (i.e., narrower confidence bands and smaller P-values) than for the full set (categories 1 to 5) – and this is the case although the sample sizes, m, are smaller. This indicates that the downward trend character for the full set (Figure 6.2a) is dominated by the lower hurricane categories, which have more events than the higher categories.

Upward trends in hurricane activity emerge when the restriction is on the heavy events. The set of events for categories 3 to 5 has $m = 92$ elements. The occurrence rate shows an overall increase (Figure 6.2d), which, however, is statistically not significant ($u_{\text{CL}} = +0.34$, $P = 0.37$). It seems that the number of 63 category-3 events "blur" the estimation. The restriction to the highest categories, 4 to 5 with $m = 29$ events, yields the clearest picture (Figure 6.2e). The Cox–Lewis test confirms the overall upward trend ($P = 0.06$).

At around 1851, the occurrence rate of events of category 4–5 is $\widehat{\lambda}(T) \approx$ 0.06 a^{-1} [0.01 a^{-1}; 0.10 a^{-1}]. One event in about 17 years. The rate steadily increased and peaked in the mid-1940s, when $\widehat{\lambda}(1944) \approx$ 0.26 a^{-1} [0.17 a^{-1}; 0.36 a^{-1}] (Figure 6.2e). This means four times more events per time unit than in 1851. Then a shorter-term downward trend is seen, which culminated in the mid-1980s, when $\widehat{\lambda}(1986) \approx$ 0.16 a^{-1} [0.09 a^{-1}; 0.24 a^{-1}]. This was a significant low in the sense that its value is below the lower bound for the estimate from 1944.

Figure 6.2e indicates that from 1986 onward, the rate of occurrence of hurricanes of category 4 or 5 again increased. This reminds us of the unallowed speculations about a similar recent upward trend behavior for the full range, categories 1 to 5, shown in Figure 6.2a. In the case of heavy events (Figure 6.2e), the recent upward trend is more pronounced but still difficult to assess due to the still wide confidence band.

From an econometric perspective, the occurrence rate curve for the heavy hurricanes (category 4 to 5) is crucial since these are the deadly and costly events. Also category 3 can mean a destructive event, such as Katrina in August 2005 (Table 6.1). Figure 6.2e shows the three major features of the curve: the long-term increase from 1851 to the mid-1940s, the low during the mid-1980s, and the recent upward trend. Could data inhomogeneities or variations in data quality be responsible for these features instead of climate variations?

First, the long-term increase in hurricane risk from 1851 to the mid-1940s. We notice the risk of missed hurricanes for the interval from 1851 to 1910 (Section 6.1.1). It is more likely to miss an event of lower category. The risk of missed smaller hurricanes should decrease over time since technological achievements allow to see better. However, the estimated occurrence rate for smaller events (categories 1 to 2) shows a long-term decrease (Figure 6.2b). This means that a time-dependent risk of missing smaller events has minimal effects on the estimation result. And the effects of a time-dependent risk of missing heavy hurricanes – if it exists – should be even smaller. We therefore conclude that the first major feature, the long-term increase in the risk of heavy hurricanes (categories 4 to 5), is robust against data inhomogeneities.

Second, the low in the risk of heavy hurricanes during the mid-1980s. We notice a possibly reduced data quality, perhaps even for the heavy events, for the interval from 1964 to 1979 (Section 6.1.1). Maybe there were more heavy events (categories 4 to 5) than listed in Table 6.1? The risk estimate for the mid-1980s, which are only about one bandwidth ($h = 20$ a) away, would certainly be influenced. We cannot rule out the possibility that the second major feature, the low in the risk of heavy hurricanes (categories 4 to 5) during the mid-1980s, is an artifact.

Third, the recent upward trend in hurricane risk. Is it true that the rate of occurrence of heavy landfalling hurricanes (categories 4–5 or 3–5) steadily increased since the low during the mid-1980s? A qualitative line of reasoning, as via the missed events for the first and second major features, is likely insufficient since the data quality for the recent decades should be good (Delgado et al. 2018). Let us therefore perform a separate risk analysis for the interval from 1980 to 2018 (Figure 6.3). The confidence band around the estimated occurrence rate curve shows a significant increase for events of categories 4 to 5, which is attested by the Cox–Lewis test (Figure 6.3b). If we include category-3 events, the test statistic is still positive, but the P-value large (Figure 6.3a). Also for the recent decades, it seems that the number of category-3 events "blur" the clear estimation picture obtained on the events of categories 4 to 5.

If we pool all events together (categories 1 to 5), then there is no significant trend for the interval from 1980 to 2018 ($P = 0.33$).

The robustness of the upward trend in the occurrence of heavy hurricanes in the recent decades (i.e., the interval from 1980 to 2018) is further examined by means of a sensitivity analysis (Section 6.1.3).

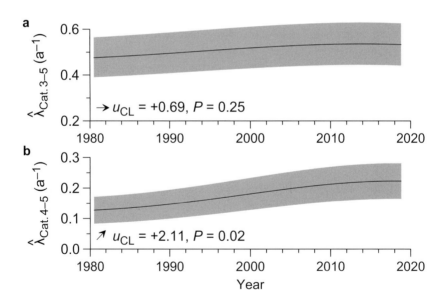

Figure 6.3 US landfalling hurricanes, 1980–2018, occurrence rate estimation. Shown are the results for various category ranges (a, 3–5; b, 4–5). The kernel bandwidth is (a, b) $h = 20$ a. The number of resamples is (a, b) $B = 2000$. For more details, see Figure 6.2.

6.1.3 Sensitivity Analysis

The third major feature of the occurrence rate curve for heavy hurricanes (categories 4–5) is the recent upward trend. For the data from the interval from 1980 to 2018, this trend is confirmed as statistically significant ($P = 0.02$) by means of the Cox–Lewis test (Figure 6.3b). The upward trend character makes the feature economically relevant.

Let us examine the robustness of the recent upward trend with respect to the accuracy and quality of the hurricane data. A hurricane is a wind-speed extreme in time and space (Section 6.1.1). It is difficult to track and measure, even today with the advanced technology. As a result, the measured wind-speed maxima, v, show uncertainties. The Atlantic Hurricane Database Reanalysis project (Delgado et al. 2018: table 2 therein) gives the following uncertainties in v for US landfalling hurricanes: interval 1954–1963, 11 kn; late 1990s, 10 kn; and late 2000s, 9 kn.

Let us overtake the uncertainty values from Delgado et al. (2018) and specify or extend them as follows. For the 1980s, we assume the uncertainty to be 11 kn, for the full 1990s, we set it as 10 kn, and for the interval up to 2018, we set it as 9 kn. For the interval from 1980 to 2018, there are 60 hurricane events (Figure 6.4): category 5 has 1 event, category 4 has 6 events, category 3 has 13 events, category 2 has 16 events, and category 1 (not shown) has 24 events.

In the case of hurricane Hugo in September 1989, the uncertainty in maximum wind speed, indicated by a Gaussian PDF (Figure 6.4), means that with a probability of $F((113-120)/11) \approx 26\%$ the real Hugo event could have been of a category less than 4 (F is the standard normal distribution function). And the real hurricane Emily in August 1993 could have been of category 4 or 5 with a probability of $F((100-113)/10) \approx 9.7\%$ (Figure 6.4).

The uncertainty in v means that a category-4 event (e.g., Hugo) could have been smaller in reality, or that a category-3 event (e.g., Emily) could have been larger. This would influence the Cox–Lewis test. The sensitivity analysis to examine the influence of the uncertainties proceeds as follows (software CLsim).

Step 1   For each of the 60 hurricanes between 1980 and 2018, generate a simulated maximum wind speed by means of a Gaussian random number, the observed v-value (Table 6.1) and the given uncertainties.

Step 2   Detect the simulated hurricane events of categories 4 or 5 as POT data for the (inclusive) threshold of 113 kn.

Step 3   Perform the Cox–Lewis test on the simulated hurricane events.

Step 4   Repeat Steps 1 to 3 until 2000 simulated P-values are available.

Step 5   Report the average of the simulated P-values.

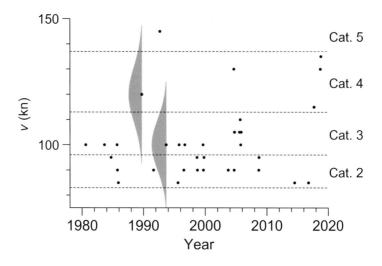

Figure 6.4 US landfalling hurricanes, 1980–2018, sensitivity analysis. Maximum wind speeds, v, for events of categories 2 to 5 are shown as filled symbols. Gaussian PDFs of v for two example events are shown as shaded areas: hurricane Hugo in September 1989 (mean 120 kn and standard deviation 11 kn) and hurricane Emily in August 1993 (mean 100 kn and standard deviation 10 kn). Category boundaries (83 kn, 96 kn, 113 kn and 137 kn) are shown as dashed lines. Cat., category.

What average P-value should we expect from the simulation exercise? If the uncertainties were zero, then we would get the original result (categories 4 to 5) for 1980–2018 (Figure 6.3b), that is, $P = 0.02$: a highly significant upward trend. If the uncertainties were very large, then the average P-value would approach the original result (categories 1 to 5) for 1980–2018 (Section 6.1.2), that is, $P = 0.33$: no significant trend. Therefore we expect that the uncertainties, greater than zero but not very large, will increase the average of the simulated P-values somewhat. The resulting average, now, is $P = 0.09$. This means a significant upward trend in the rate of occurrence of heavy hurricanes (categories 4 to 5) during the interval from 1980 to 2018. We therefore conclude that the third major feature, the recent upward trend, is robust against the presence of wind-speed uncertainties.

6.2 Case Study: Paleo Hurricanes

Although a hurricane originates in the tropical North Atlantic–West Indies region, the cyclonic low-pressure system can move in the northern direction

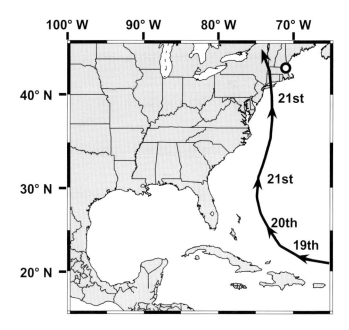

Figure 6.5 Great New England Hurricane. This category-3 event ($v = 105$ kn) made landfall at Long Island (New York) and Connecticut on 21 September 1938. The shown path is drawn after www.nhc.noaa.gov/outreach/history/#new (17 June 2019). Also shown is the location of Boston (open circle).

and hit mid-latitude land areas, such as New England in the United States. One event was the Great New England Hurricane in September 1938, which is listed in the database (Table 6.1). It hit the states New York, Connecticut, Rhode Island, and Massachusetts (Figure 6.5). There were wind-speed, air-pressure, and temperature observations available (Pierce 1939), and the hurricane was of category 3.

According to the National Hurricane Center of the United States, the Great New England Hurricane cost a number of approximately 600 lives and made overall losses of more than 300 million USD (www.nhc.noaa.gov/outreach/ history/#new, 17 June 2019). Taking into account inflation and changes in wealth, the total economic damage of this event, normalized for the year 2005, would have been nearly 40 billion USD (Pielke et al. 2008).

The Lower Mystic Lake near Boston, Massachusetts provides a sedimentary archive of past hurricane events (Section B.2). Extreme thickness of an annual varve, in combination with a graded bed, indicates a hurricane (Besonen et al. 2008). The parameters for the detection of the extremes (background

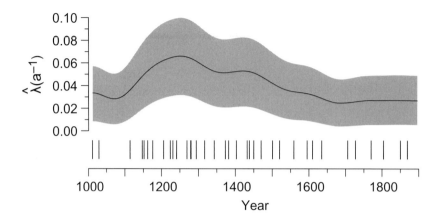

Figure 6.6 Hurricanes near Boston, AD 1011–1897, occurrence rate estimation.
The number of events of this proxy series from Lower Mystic Lake is $m = 36$.
The occurrence rate (solid line) with 90% confidence band (shaded) is estimated
by means of a Gaussian kernel with a bandwidth of $h = 50$ a. The number of
resamples is 2000.

smoothing and threshold) are tuned such that a good agreement with documentary information (available from about 1630) is obtained (Figure 2.6).
The varve thickness record from Lower Mystic Lake allows to extend the
hurricane record back to AD 1011. The end of the observation interval is
1897 – there are no varves, no proxy information in the later time interval
due to changes in the hydraulic regime of the lake (Section 2.4.1). There are
36 detected hurricane events near Boston (Figure 6.6).

6.2.1 Occurrence Rate Estimation

Boundary bias correction via the reflection method of pseudodata generation
(Section 3.2.3) is relevant especially for the lower observation-interval boundary. There is one extreme event in the first year of the record (AD 1011). Since
the reflection method would generate a pseudodata point at the same year,
we set the start of the observation interval to AD 1010. The cross-validation
function (not shown) does not exhibit a well-defined local minimum and is not
helpful for guiding bandwidth selection. For the purpose of the selection of
the bandwidth, h, we play and find out by means of the confidence band that
the selection of a value of $h = 50$ a yields a good solution to the smoothing
problem for the paleo hurricane data. The number of resamples ($B = 2000$)
follows the usual recommendation (Section 3.2.4).

The estimation result (Figure 6.6) reveals as major feature of the occurrence rate a well-expressed high during the thirteenth century. The peak value with 90% CI is $\widehat{\lambda}(T) \approx 0.07$ a^{-1} [0.03 a^{-1}; 0.10 a^{-1}]. Before and after the peak, the rate was significantly lower, down to a value in the order of 0.03 a^{-1}. While the trend before (eleventh to twelfth century) was stronger upward, it was more gradual downward after the peak (fourteenth to nineteenth century). The Cox–Lewis test does not reflect a significant overall trend ($u = -1.15, P = 0.12$). This is as expected, since the test, performed on the full interval, is influenced by the upward–downward local trends. This illustrates the limitations of the test.

How does the result on the paleo hurricanes near Boston compare with the findings obtained on US landfalling events (Section 6.1)? First, the overall occurrence rates for the Boston area (Figure 6.6) are clearly smaller than the rates for US landfalls (Figure 6.2), even if only the heavy events (categories 4 to 5) are considered. This makes sense since the majority of landfalling hurricanes miss the Boston area while on their way in the United States.

Second, let us consider the overlapping period between the paleo record and the observed landfalls, which is the interval from 1851 to 1897. The paleo record shows here just one hurricane event, namely for the year 1869. The observations (Table 6.1) inform that this was the Eastern New England event from 8 September 1869, which was of category 3. The metadata (www .aoml.noaa.gov/hrd/hurdat/Data_Storm.html, 6 June 2019) list, which federal state was affected at which category. There are another four events for Massachusetts in the database (Table 6.1), but not in the paleo record: New England on 16 September 1858, Saxby's Gale on 4 October 1869, an unnamed event on 18 August 1879 (which came to Massachusetts one day later), and another unnamed event on 10 September 1896. However, each of the four events was only of category 1 when it hit Massachusetts. The failure by the paleo hurricane archive to detect those four events can be considered a minor shortcoming.

6.3 Case Study: Storms at Potsdam

The strongest storm on record at Potsdam, Germany, was the "Niedersachsen-Orkan" on 13 November 1972 with a maximum wind speed of 28.4 m s^{-1} (Figure 6.7). In the spatial domain, this event was strongest at the German Bight region, roughly 400 km away from Potsdam, and it caused 47 fatalities in Germany (Deutscher Wetterdienst 2002). Big winter storms at Potsdam

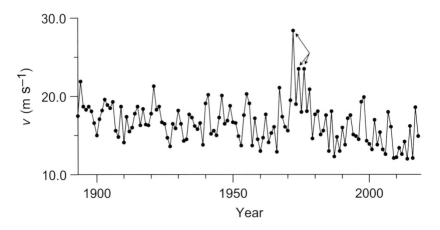

Figure 6.7 Annual block maxima, maximum daily wind speed at Potsdam. Three prominent events during the 1970s (13 November 1972, 29 December 1974, and 3 January 1976) are indicated (arrows).

occurred also on 29 December 1974 and 3 January 1976 (Figure 6.7). The costliest winter storm so far in Germany was "Kyrill" on 18–19 January 2007 with 13 fatalities and not-inflation-adjusted overall losses of 5.5 billion USD according to Munich Re (see Reading Material). This event was also recorded by the Potsdam series (Figure 2.3) with a maximum wind speed of 18.0 m s^{-1}.

A winter storm in Germany is more a manifestation of a cyclone, that means, it has large horizontal spatial dimensions, up to thousand kilometers and more (Pinto and Reyers 2017). A convective storm, on the other hand, results from thermally driven vertical air motions, and it has comparably small dimensions (Allen 2018). It is also called thunderstorm. One example, the tornado, is a rotating column of air that is generally less than 500 m in diameter and has very strong winds. In Germany, the big convective storms occur in summer. Convective storms may be costly, since they may also be associated with hail or heavy rainfall. The costliest event so far in Germany was on 27 to 28 July 2013 with not-inflation-adjusted overall losses of 4.8 billion USD according to Munich Re. However, this event was not recorded by the Potsdam series (Figure 2.3), where the maximum wind speed during these days was just 6.5 m s^{-1}. Also the next four costliest events were "missed." This documents the localized nature of summer storms.

The IPCC concluded that there are considerable uncertainties about long-term changes of the statistical properties of extreme extratropical cyclones (Hartmann et al. 2013: section 2.6.4 therein). Different types of data yield partly conflicting results, and long homogeneous wind-speed records are sparse. The situation has not much improved today (Vautard et al. 2019). The Potsdam series of maximum daily wind speed can shed some light on the issue of long-term trends in storm occurrence. We start with a stationary analysis framework and then consider nonstationarity.

6.3.1 Stationarity

Let us assume that the PDF of the wind speed at Potsdam has not changed over time. What is the return period for an event as strong as the Niedersachsen-Orkan? The annual block maxima values (Figure 6.7) are calculated from maximum daily wind-speed data, which itself are obtained from 24 hourly mean values. The approximation of the distribution of block maxima by a GEV distribution (Section 3.1.1) should be excellent since a block maximum is taken from many (on average $24 \times 365.25 = 8766$) values.

The GEV parameters are estimated by means of the maximum likelihood technique (software GEVMLEST) as $\widehat{\mu} = 15.4 \pm 0.2 \text{ m s}^{-1}$, $\widehat{\sigma} = 2.3 \pm 0.2$ m s^{-1}, and $\widehat{\xi} = -0.07 \pm 0.05$. The latter estimate is clearly larger than -0.5, which indicates that the regularity conditions for a successful estimation are fulfilled. The histogram of the block maxima roughly agrees with the fitted GEV distribution (Figure 6.8a).

The increase of the return level with the return period is displayed in Figure 6.8b. The shown standard error bars are scaled with a factor of 2.3264,

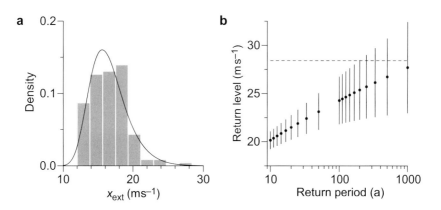

Figure 6.8 Annual block maxima, maximum daily wind speed at Potsdam, GEV estimation. (a) Estimated GEV density function (solid line) and histogram estimate of empirical density. (b) Return level with standard errors (multiplied by a factor of 2.3264) for the fitted GEV model in dependence on the return period; see the text for explanation. The horizontal dashed line is at $v = 28.4 \text{ m s}^{-1}$.

which equals approximately the percentage point of the normal distribution function at 99% (Mudelsee 2014). In other words, what exceeds an estimated return level plus the scaled bar length, can be assigned a return period of at least the corresponding value at the 99% confidence level. The Niedersachsen-Orkan, which had $v = 28.4$ m s^{-1}, was just not a 200-year event, but it was at least a 167-year event (Figure 6.8b). We pursue the analysis of this and other events in the nonstationary analysis framework.

6.3.2 Nonstationarity

Let us now test whether the PDF of the wind speed at Potsdam has changed over time. A basic exercise is to fit a straight line to the series of annual block maxima (Figure 6.9). The fit shows unambiguously a negative slope of $\widehat{\beta}_1 = -0.022 \pm 0.008$ m s^{-1} per year. The error bar is determined via bootstrap resampling. That means, the downward trend is a robust finding.

This finding does not invalidate the GEV analysis from Section 6.3.1. The negative trend direction of the annual maxima can be considered as a basic step to fit a GEV model with linearly time-dependent location parameter, $\mu(T)$ (Eq. 3.13). Also the other GEV parameters, σ and ξ, can be made time-dependent. The technical challenge then is to obtain numerically a reliable estimation (Section E.1).

However, the major conclusion to be drawn from the downward trend in annual wind-speed maxima regards the Niedersachsen-Orkan. This event, on 13 November 1972, happened well in the second half of the observation

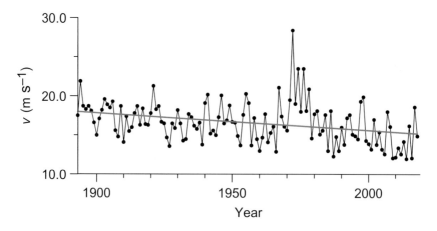

Figure 6.9 Annual block maxima, maximum daily wind speed at Potsdam, linear trend estimation. The trend (gray line) is fitted by means of OLS (Section D.1.2).

interval (1893–2018), when the long-term tendency for annual maxima was already reduced, by approximately (79 a) $\times \widehat{\beta_1} \approx 1.7 \pm 0.6$ m s^{-1}. Thus, it appears even more surprising that this event happened. More technically, the true return period (i.e., the inverse of the time-dependent tail probability) for the Niedersachsen-Orkan was likely greater (i.e., 200 years or more) than on basis of the stationary analysis (Section 6.3.1).

More relevant from a risk-analytical perspective is occurrence rate estimation (Section 3.2.1) combined with a sensitivity analysis of the effects of the selection of the wind-speed threshold. This methodology helps to answer questions after the time-dependent storm risk in a quantitative manner. It is more relevant for the practitioner to be able to quantify the probability that an event of a certain minimum size occurs within a time interval than to evaluate the size of annual maxima.

Boundary bias correction via the reflection method (Section 3.2.3) is somewhat relevant for the lower boundary since there, at the end of the nineteenth century, some minor storms occurred (Figure 6.10a). At the upper boundary, toward the present, not many events are recorded. The cross-validation function (not shown) indicates usage of a kernel bandwidth of $h = 1.6$ a. However, the bootstrap confidence band (not shown) renders the many wiggles in the occurrence rate, which can be seen via such a small bandwidth, as insignificant. Therefore we play with the bandwidth and increase it. We find that in the case of a moderate upper threshold (8 Beaufort), a value of $h = 10$ a yields a better solution to the smoothing problem (Section 3.2.2). This bandwidth selection allows structure in the occurrence rate to be inspected that is still significant (Figure 6.10a). In the case of major events (9 Beaufort and above), the data sparseness does not permit such a detailed inspection, and we set $h = 30$ a (Figure 6.10b). The number of resamples ($B = 2000$) follows the usual recommendation (Section 3.2.4).

For a moderate upper threshold (8 Beaufort), there was a high in storm risk at Potsdam from the start of the observations in 1893. The occurrence rate then was $\widehat{\lambda}(T) \approx 1.25$ a^{-1}, that means, there happened on average more than one event per year (Figure 6.10a). A gradual decrease set in, and during the 1940s, the occurrence rate was about half of the value at around 1893. Then a relatively fast increase set in. Note that owing to the nonzero bandwidth, the variations – especially the fast ones – are smoothed over in the estimation. A peak in the risk of storms (of 8 Beaufort and higher) was reached in the 1970s, when again $\widehat{\lambda}(T) \approx 1.25$ a^{-1} (Figure 6.10a). Since then, a steady decrease in storm risk at Potsdam is observed. The estimate for the end of the observation interval (31 December 2018) is $\widehat{\lambda}(T) \approx 0.14$ a^{-1}, for which the 90% CI equals [0.05 a^{-1}; 0.25 a^{-1}].

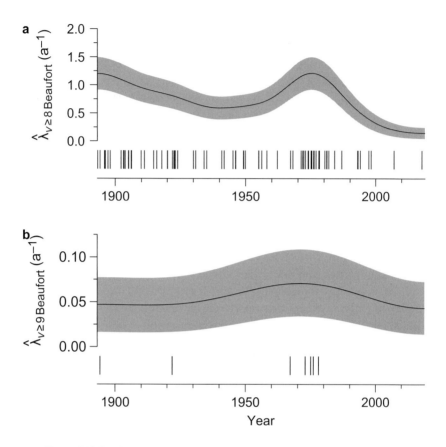

Figure 6.10 Maximum daily wind speed at Potsdam, occurrence rate estimation. Shown are the results for two threshold settings (a, ≥ 8 Beaufort or 17.2 m s^{-1}; b, ≥ 9 Beaufort or 20.8 m s^{-1}). The number of events is (a) $m = 93$ and (b) $m = 7$. The occurrence rates (solid lines) with 90% confidence bands (shaded) are estimated by means of a Gaussian kernel with a bandwidth of (a) $h = 10$ a and (b) $h = 30$ a. The number of resamples is 2000.

For major storms (9 Beaufort and above), it is close to impossible to demonstrate significant time-dependent variations in occurrence rate (Figure 6.10b). This is owing to the sparseness of such events. The occurrence rate is $\widehat{\lambda}(T) \approx 0.05$ a^{-1}, that means, one event per 20 years. There are indeed some events located in the 1970s, such as the Niedersachsen-Orkan, but they are too few to make a peak in storm risk detectable.

We have two major statistical findings on storm risk at Potsdam during the instrumental period (Figure 6.10). These are (1) the long-term downward trend that was (2) interrupted by a local high during the 1970s, when the occurrence

Table 6.2. *Maximum daily wind speed at Potsdam, Cox–Lewis test. The test (Section 3.2.5) of the null hypothesis "constant occurrence rate" (no trend) is performed in dependence on the upper, inclusive threshold,* u. *The analysis is performed separately for the full year and the winter season (November to February). A single test result is obtained on a sample of size* m *of extreme events and is given as test statistic (*u_{CL}*) together with the one-sided* P-*value.*

Threshold	Season					
	Full year			Winter		
u (m s^{-1})	m	u_{CL}	P	m	u_{CL}	P
17.2	93	−2.67	0.004	62	−2.21	0.01
17.6	77	−2.39	0.008	49	−2.22	0.01
18.0	54	−1.84	0.003	32	−2.21	0.01
18.4	37	−1.34	0.09	25	−1.93	0.03
18.5	36	−1.18	0.12	20	−1.94	0.03
19.0	20	0.02	0.49	10	−0.63	0.26
20.0	10	−0.28	0.39	8	−0.08	0.47
20.8	7	−0.07	0.47	7	−0.07	0.47

rate reached values comparable to those at around the end of the nineteenth century. To test harder these findings and raise the robustness of the conclusions, let us play with the parameters and study the sensitivity of the results.

First, the long-term downward trend. This feature is clearly visible when the threshold is set to the moderate value of 17.2 m s^{-1} or 8 Beaufort (Figure 6.10a), but the trend is not there when the threshold is set to the larger value of 20.8 m s^{-1} or 9 Beaufort (Figure 6.10b). What is with the wind-speed thresholds in between? Table 6.2 shows the results of the Cox–Lewis test for various upper thresholds between 17.2 and 20.8 m s^{-1}. The test is performed separately for the full year and the winter season. The test results basically confirm the first major statistical finding, the existence of a long-term downward trend. This is detectable at the 90% confidence level as long as the upper threshold, u, is small enough (u = 17.2 to 18.4 m s^{-1}). When u is increased further, then too few events, m, jump above, and the P-value of the test increases, in other words: no detectability.

If we focus on winter storms (here defined as from November to February), then the Cox–Lewis test yields basically the same results as for the full year (Table 6.2). The table also shows that most of the storms at Potsdam are winter storms. The focus on storms during the summer season (from May to August) does not yield enough events (m = 2 for u = 17.2 m s^{-1}) for a statistical analysis.

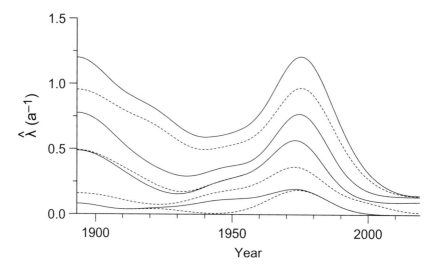

Figure 6.11 Maximum daily wind speed at Potsdam, occurrence rate estimation, sensitivity analysis. The storm occurrence rate (alternating solid and dashed lines) is estimated for upper thresholds, u, equal to (from top to bottom) 17.2, 17.6, 18.0, 18.4, 18.5, 19.0, 20.0, and 20.8 m s^{-1}. The other estimation parameters are $h = 10$ a and $B = 2000$.

Second, the local high in storm occurrence during the 1970s. The sensitivity analysis studies the effects of the threshold selection on the occurrence rate curve. The selection of $u = 17.2$ m s^{-1} or 8 Beaufort results in the topmost curve in Figure 6.11 (for this curve together with the confidence band, see Figure 6.10a). The increase of u leads to a reduced number, m, of detected events and to generally reduced occurrence rates (Figure 6.11). However, the general shape of the curve – a long-term downward trend on which a local high during the 1970s sits – remains "intact" when the threshold is increased over $u = 17.6$ m s^{-1} and further to up, say, $u = 19.0$ m s^{-1} (the third curve from the bottom in Figure 6.11). Also the second major statistical finding – the local high in storm occurrence during the 1970s – enjoys robustness over a large range of selected thresholds.

Note that it is legitimate to compare the occurrence rate curves in Figure 6.11 without consultation of the confidence band. This is because the uncertainties of the curves are strongly mutually dependent on each other. In other words, if the true (but unknown) occurrence rate of certain storms (e.g., equal to or larger than $u = 17.2$ m s^{-1}) in a certain year, say 1962, would be higher than the estimate, then also the true rate for heavier storms (e.g., equal to or larger than $u = 17.6$ m s^{-1}) in 1962 would very likely be higher than the corresponding estimate. This strong dependence is visualized as well in Figure 6.11.

Let us briefly summarize the case study on storms at Potsdam, Germany. The first major finding – the long-term downward trend in storm risk during the instrumental period – is compatible with what is reported by the IPCC on the basis of observations at other mid-latitude locations (Hartmann et al. 2013: section 2.6.4 therein). However, the trend for Potsdam is shown as a highly robust finding obtained on excellent, homogeneous data. This quality allows to obtain the second major finding – the local high in storm occurrence during the 1970s. Also this result is shown to be robust by means of a sensitivity analysis. This is an intriguing observation, since it seems that such a local high has not been previously reported for this or other sites. The largest event at Postdam, a winter storm called the Niedersachsen-Orkan on 13 November 1972, happened during that local high. What are the climatological causes of the downward trend and the local high during the 1970s? What are the meteorological causes of the Niedersachsen-Orkan? How are the climate and weather aspects associated here?

6.4 Outlook

This chapter is on positive extremes (at the high end) of the climate variable wind speed. We have applied the statistical methods (Chapter 3) to two types of storms: hurricanes and extratropical cyclones. Let us assess separately for the two types how the case studies can be extended in the three research directions space, time, and resolution.

Hurricanes originate in the tropical North Atlantic–West Indies region. The first case study (Section 6.1) is on observed landfalling events in the area of the United States during the interval from 1851 to 2018. The second case study (Section 6.2) is a spotlight on the region near Boston for the interval from AD 1011 to 1897, which is obtained by means of a natural climate archive (lake sediment core). This second case study can be called an extension of the first study in the research direction time. And the first study can be called an extension of the second study in the research direction space.

The mantra for more data to be analyzed for achieving a better understanding applies also to the study of hurricanes. More paleoclimate archives have to be tapped, and more proxy variables for high wind speeds have to be developed and measured, in order to go further back in time. As an example, Mann et al. (2009) employ overwashed sedimentary deposits from various coastal sites in the United States to reconstruct the hurricane activity over the past 1500 years.

Mann et al. (2009: p. 880 therein) find a "peak in Atlantic tropical cyclone activity during medieval times (around AD 1000)," which may be compared with the peak during the thirteenth century found in the second case study (Figure 6.6). Taking into account the dating uncertainties of the sedimentary deposits, which are in the order of several decades (Mann et al. 2009: figure 1 therein), and the analysis bandwidths, which are 50 years for the second case study and 40 years for the deposits (Mann et al. 2009: figure 3 therein), it is not straightforward to reject the synchroneity hypothesis.

However, many more paleo spots have to be sampled and analyzed in order to come closer to the quality of the observed US landfalling hurricane database (Table 6.1) in terms of spatial coverage. The early part of the database, back to 1851, is based also on historical documents. Such data can be used to extend the hurricane history in time back to the discovery of the New World in 1492, as has been done for the deadliest events (Rappaport and Fernández-Partagás 1997).

One further advantage of documentary hurricane data is the potential availability of the day and month of an event in a year. This allows to identify several events in a season. Such a differentiability and time resolution is almost impossible to achieve for hurricane proxy data from natural archives.

The second storm type considered in this chapter are those events that originate outside of the tropics. The case study on storms at Potsdam in Germany (Section 6.3) includes (1) winter storms, which have large dimensions (up to thousand kilometers and more) of horizontal, cyclonic air flow and (2) summer storms, which are smaller, thermally driven vertical motions.

The extension of the Potsdam series, which is from 1893 to 2018, backward in the research direction time – it seems impossible. There are no other weather stations nearby with series of that length. In fact, it seems fair to say that there are few other meteorological series from elsewhere in the world that go back as far as Potsdam and that are of such a good data homogeneity (Section 2.2.1).

The extension in the research direction space is possible if the condition on the start of the observation interval is relaxed. The past few decades should already allow an assessment of climatic (i.e., long-term) influences on short-term storminess at mid-latitudes. For Potsdam, we find a long-term downward trend in storminess (Figure 6.10a). There exist analyses of wind observations from other stations in Europe, which also report long-term downward trends. Examples are the papers by Schiesser et al. (1997) on winter storms in Switzerland from 1864 to 1994 and Smits et al. (2005) on storms in the Netherlands over the full year from 1962 to 2002. However, there are also other papers. Hammond (1990) on storms in Britain from 1881 to 1990 was unable to find a clear trend. Pirazzoli and Tomasin (2003) on storms in the Mediterranean–Adriatic region from 1951–2000 find indications for a trend change (from downward to upward) during the 1970s. For Potsdam, we find as second major feature the local high in storm risk during the 1970s (Figure 6.10a). This is a robust finding. There exist no indications in the research literature for a similarly strong local high elsewhere.

The situation of observational data from the past decades is such that researchers are prevented from detailed learning about longer-term changes in storminess over larger geographical areas at high resolutions in space and time. This verdict applies to hurricanes and extratropical storms, but especially to smaller-scale phenomena. There is low confidence in observed long-term trends in the occurrence of thunderstorms in summer and the associated effects, such as hail or flashfloods (Hartmann et al. 2013: p. 216 therein). We are more optimistic as regards the detailed learning about shorter-term changes because of the improved observation systems (satellites), such as the Global Earth Observation System of Systems (GEOSS). Such systems should improve the situation of shorter-term data in terms of spatial coverage and resolutions in space and time.

A hurricane forms over the ocean, moves, makes landfall or not, and, finally, decays. The category assessments, born out of the necessity to give a single number, are inevitably uncertain (Section 6.1.1). Therefore it is right that there are only a handful of different categories. With the improved observation systems, it becomes theoretically possible to track a hurricane as a wind-speed extreme in space and time. It appears obvious that heuristic pattern-recognition algorithms ("artificial intelligence") from numerical mathematics (Appendix E) will become even more important in the future to help us researchers to make practically feasible the recognition and tracking of hurricanes – and other storms as well.

6.5 Summary for the Risk Analyst

Economically relevant are the current situation and the near future (the next five years). For strategic decisions, also the next few decades count. The ideal resolution of wind-speed input data for making risk analyses on this time range is daily or better (i.e., smaller). Note that the input data are obtained from very many wind-speed values per day. For example, hurricanes (Section 6.1) are categorized via the highest sustained one-minute wind speed, and storms in Germany (Section 6.3) via the highest sustained ten-minute wind-speed. A certain number of geographical sites has to be studied to achieve a comprehensive storm risk analysis for a larger region.

However, it is possible to learn also from univariate series, namely about those storm types that affect a wider area. This is shown by the case studies on US landfalling hurricanes (Sections 6.1 and 6.2) or storms in Potsdam in Germany (Section 6.3). With the tendency toward improved observation systems (i.e., satellites), which generate many data points in space and time, the recognition and tracking of hurricanes and other storms will become feasible. The achievement of this task requires to adapt the risk-analytical methods of detection and estimation (Chapter 3) into high-dimensional, nonstationary analysis frameworks.

Storm risk, in the tropical region (e.g., hurricanes) and elsewhere, is not well understood. One reason are the strong spatial and temporal variations of the climate variable wind speed. The available data do not allow sufficiently well to capture the climate variability. Simplified meteorological arguments do not help in this situation. Climate warming affects the Arctic and Antarctic stronger than other regions (polar amplification). As a result, the equator–pole temperature gradient, and consequently the overall air-pressure differences, are likely to get smaller in the future. Therefore fewer storms at mid latitudes? The physical reality, however, is more complex (Cohen et al. 2014).

It is the magnitude that determines the impacts of a storm. This allows the risk analysis to focus on wind-speed maxima. The duration aspect is less relevant, and there is no need to form index variables. This simplifies the analytical task.

The risk of occurrence of heavy hurricanes (categories 4 to 5) has significantly increased during the interval from 1980 to 2018 (Figure 6.3b). A sensitivity analysis confirms the robustness of this result against data uncertainties (Section 6.1.3). The inclusion of weaker hurricanes (category 3) "blurs" the estimation picture. This upward trend is a finding obtained by means of advanced statistical methods (Chapter 3). It confirms previous considerations (Walsh et al. 2016). It should have socioeconomic relevance.

For Potsdam in Germany, the storm level relevant for the insurance industry is 8 Beaufort. The first major statistical finding of the case study (Section 6.3) is the existence of a significant downward trend over the full observation interval from 1893 to 2018 (Figure 6.10a). A sensitivity analysis confirms the robustness of this result against threshold selection (Table 6.2). The second major finding is the local high in storm occurrence during the 1970s. The storm risk in the immediate decades before and after was significantly smaller, as the confidence band (Figure 6.10a) and the sensitivity analysis (Figure 6.11) attest. It seems that such a local high has not been previously reported for this or other sites. The largest event recorded at Potsdam was the Niedersachsen-Orkan on

13 November 1972. This event has a return period likely greater than 150 or even 200 years (Section 6.3.1).

For the researcher of extratropical storm risk, there are several open questions on the long-term downward trend observed at Potsdam, the local high during the 1970s, and the occurrence of the Niedersachsen-Orkan during that decade. What are the climatological causes? How are the climate and short-term weather aspects associated? Regional climate models (Appendix C) are not yet helpful here. The task of predicting the time-dependent occurrence rate into the near future is therefore likely best performed by means of extrapolation of the curve.

Also for the analyst of hurricane risk, due to the lack of knowledge about the driving factors (Walsh et al. 2016), risk prediction by means of coupled climate models – is not yet feasible. This is also reflected in the Fifth Assessment Report of the IPCC, namely in the chapters on climate projections and predictability (Kirtman et al. 2013: p. 956 therein) and on the relevance of climate phenomena for future regional climate change (Christensen et al. 2013: section 14.6.1 therein). The coupled models are at the moment useful for the study of what-if questions and strategic development, not for risk quantification.

Reading Material

The National Hurricane Center of the United States published a comprehensive report about hurricane Katrina (Knabb et al. 2011) at www.nhc.noaa.gov/data/tcr/AL122005_Katrina.pdf (21 May 2019) and issues a regularly updated list of the costliest US hurricane events, which is available at www.nhc.noaa.gov/news/UpdatedCostliest.pdf (5 June 2019). Pierce (1939) is a contemporary review of the meteorological history of the Great New England Hurricane of 1938. The reinsurance company Munich Re provides online catalogues of recent costly natural catastrophes, including storms, at the internet site www.munichre.com/de/reinsurance/business/non-life/natcatservice (22 May 2019). Weinkle et al. (2018) review hurricane damages in the United States using a normalization that takes into account inflation and other changes in societal conditions.

The Glossary of Geology (Neuendorf et al. 2005) explains many expressions from the Earth sciences. The Glossary of Meteorology by the American Meteorological Society can be accessed online at http://glossary.ametsoc.org (31 May 2019).

GEOSS data can be accessed via www.geoportal.org (21 June 2019).

Emanuel (1987) explains the physics of hurricanes and that they "make textbook examples of the operation of a Carnot engine." Elsner and Kara (1999) is an accessible book on hurricanes of the North Atlantic (including US landfalling).

Giffard-Roisin et al. (2018) is a conference paper, available at the internet site https://hal.archives-ouvertes.fr/hal-01851001/document/ (27 June 2019), on hurricane track prediction by means of a neural network, which is a special type of pattern-recognition tool.

Ting et al. (2019) is an article that studies the vertical wind stress – a metric that controls Atlantic hurricanes. On the basis of climate model simulations from the Coupled Model Intercomparison Project Phase 5, Ting et al. (2019) predict in the abstract of their paper that during the twenty-first century the anthropogenic GHG forcing will degrade the natural wind-stress barrier, which "allows hurricanes approaching the US coast to intensify more rapidly."

Polar amplification means that temperature changes at high latitudes exceed those at low latitudes – as has been observed for the recent decades (Cohen et al. 2014) and the paleoclimate (Masson-Delmotte et al. 2013: box 5.1 therein). The simple assumption would be that the equator–pole temperature gradients decrease with climatic warming, which would reduce the storm risk at mid-latitudes. The physical reality, however, is more complex. Cohen et al. (2014) review three potential dynamical pathways that link Arctic amplification to northern extreme mid-latitude weather: changes in storm tracks, jet streams, and planetary waves. They conclude that large uncertainties remain.

Appendix A Climate Measurements

This book studies extremes of the climate variables precipitation, temperature, and wind speed. The numerical values (denoted as small x) for these variables are obtained by measurements. It is preferable to have available an instrument for direct observation of the variable of interest since this is more accurate than indirectly measuring a proxy variable, that means, an indicator variable. For example, direct temperature observations have been made since the invention of the thermometer in the seventeenth century; earlier temperatures have to be inferred from an older climate archive (Appendix B) that stores temperature information.

A.1 Temperature

The longest instrumental temperature time series is that from central England (Manley 1974). It goes back to the year 1659. Other places followed, and by the start of what is called the instrumental period, in 1850, a network of land surface-air temperature observatories had emerged. Potsdam (Figure 1.2) opened in 1893. Today, temperature is directly observed not only on the land surface but also at the surface and in the depth of the sea by means of ships or floats, and also up to the stratosphere by means of balloons or satellites. For these modern observational data, the issues of spatial representativeness and homogeneity (Exercise 2.2) have to be addressed as well. It is regrettable that in the past years national weather services have been abandoning man-operated observations, which damages the continuity and homogeneity of climate time series.

A frost ring in a tree (Section 2.1) is a rather crude proxy for cold temperatures, but this information may date further back in time than thermometer records. Tree-rings are an example of a climate archive (Appendix B). There are many proxy–archive combinations that are employed for inferring past temperatures. The major issue with proxy variables, in general, is that they are influenced not only by the climate variable they are hoped to indicate but also by other variables. For example, in paleoclimatology, $\delta^{18}O$ measured on microfossil samples from the seafloor archive (see Section A.2 for the delta notation) records not only the temperature of the water in which the microfossils lived but also the $\delta^{18}O$ of the water, which itself is influenced by global ice volume. Evidently, the timescales of the variations play a role for assessing proxy time series. In the paleoclimate example, inferred decadal-scale temperature fluctuations would hardly be disturbed by ice-volume variations, which act on longer timescales.

A.2 Precipitation

The precipitation collector (rain, snow, and ice) developed by Gustav Hellmann (1854–1939) is still a widely used instrument. Like the thermometer (Section A.1), it delivers a point measurement. The precipitation over larger areas is today determined by means of radars, which measure the (precipitation-dependent) reflection of electromagnetic pulses sent into the atmosphere. As with any instrument, devices for measuring precipitation are not perfect; they show uncertainties. One error source for radar-measured precipitation is called clutter, that is, pulse-reflecting sources other than precipitation.

Stalagmites, which are fed by dripping water in a cave, grew long before the instrumental period. The very existence of a stalagmite is therefore already a crude proxy for past precipitation. The speleothem archive in general, to which stalagmites belong, offers the possibility to measure the isotopic composition of the material in order to obtain a more accurate proxy for precipitation (Figure 1.3). Typically measured are the oxygen isotopes ^{16}O and ^{18}O. The values are given in delta notation: $\delta^{18}O = \left[(^{18}O/^{16}O)_{sample}/(^{18}O/^{16}O)_{VPDB} - 1 \right] \cdot 1000\permil$, where $(^{18}O/^{16}O)$ is the number ratio of the isotopes and VPDB is the "Vienna Pee Dee Belemnite" standard. This formula basically tells you by how many per mil a sample's isotope ratio is larger or smaller than the ratio of the standard.

Carbon isotopes, $\delta^{13}C = \left[(^{13}C/^{12}C)_{sample}/(^{13}C/^{12}C)_{VPDB} - 1 \right] \cdot 1000\permil$, are also measured on stalagmites and can be used as a proxy for precipitation and other influences, such as vegetation.

It can be said that obtaining paleo-precipitation values over space and time via proxy–archive combinations is a harder challenge than obtaining paleo-temperatures (Section A.1). One major source of proxy error of $\delta^{18}O$ in a speleothem is that temperature fluctuations in the cave may influence the isotopic value. What helps is cave monitoring in the present and assuming that what is monitored (e.g., the seasonal temperature cycle) is valid also for the distant past. Evidently, paleoclimatology cannot proceed without invoking such an assumption of actualism.

Documentary collections constitute an archive that supplies direct point data, such as Hellmann's collector. However, because they are based on human qualitative assessment rather than on a scientific instrument, such data are clearly more uncertain; see Section 2.1 on the construction of the Elbe flood record.

A.3 Wind Speed

The anemometer for measuring wind speed via the rotation of wind-driven rotating cups has changed little since its development by Leon Battista Alberti (1404–1472). Also the Potsdam series (Figure 2.3) was obtained by means of this instrument. The anemometer delivers a point measurement. Since wind speed varies spatially much more than many other variables, for example, temperature, the point value is less representative for the area. This hampers the comparison of point measurement data and climate model data, which are output for geographical grid boxes (Appendix C). There is a tendency toward improved observation systems via the inclusion of remote-sensing devices installed on satellites (see Reading Material).

For the paleo-world, without instruments, it is impossible to obtain quantitative wind-speed data in a manner similar to the anemometer. We have to resort to rough assessments, such as whether a storm occurred at a certain place and date. Such assessments can be found in documentary data and in natural climate archives (Appendix B). The paleo-hurricane time series studied in this book (Figure 2.6) stems from a lake sediment core, with the employed proxy variable being varve thickness. Proxy variables means proxy error. Confounding variables (other than wind speed) may have influenced the varve thickness series, such as sediment flux. However, Besonen et al. (2008) assessed such disturbances as minor and the proxy quality here can be considered as good.

Reading Material

Manley (1974) is a paper on surface-air temperature, which includes a table of average monthly values from January 1659 to December 1973. It also discusses the representativeness of point measurements and homogeneity. Mudelsee et al. (2014) is a paper on $\delta^{18}O$ from the seafloor archive. It also discusses the influences of temperature and ice-volume variations on the proxy.

The European Space Agency released the satellite Aeolus into orbit on 22 August 2018. Aeolus carries a Light Detection and Ranging (LIDAR) system for the measurement of wind speed. This system provides wind-speed data up to an altitude of 2 km with an uncertainty of $1\,\mathrm{m\,s}^{-1}$ and determines the average wind speed over tracks of approximately 100 km length (www.esa.int/ Our_Activities/Observing_the_Earth/Aeolus/Measuring_wind/, 25 June 2019).

Appendix B   Natural Climate Archives

Climate archives store information about past climates. Utilizing the archives helps to look further back in time than what is possible with measurement instruments. Natural archives, such as speleothems, lake sediment cores, or tree-rings, accumulate their material over time: they grow. The age–depth or growth curve transforms the depth values in an archive, where samples are taken for climate proxy measurement (Appendix A), into age or time values (denoted as small t). Man-made historical archives may be consulted as well for the construction of a paleoclimate time series; these are not covered here but in Section 2.1.

B.1 Speleothems

Speleothems are calcium carbonate rocks that are formed in caves. Stalagmites are the speleothem type most important for paleoclimatology. The carbonate material enables the measurement of $\delta^{18}O$ and $\delta^{13}C$, which are proxies for precipitation and other influences (Section A.2). Stalagmites sit on the floor of a cave. They are fed by dripping water and accumulate the material in layers; the old material is at the bottom, and the young is at the top of the stalagmite (Figure B.1).

The carbonate material of a speleothem usually contains elemental impurities, of which uranium is important since it can be used for absolute datings (Section B.5). Such accurate time information has been recognized as a major advantage since speleothems became fashionable as climate archives in the 1990s. On the other hand, the challenge associated with speleothems is the interpretation of the proxy signal. For example, $\delta^{13}C$ variations may reflect a variety of inorganic and organic influences, which are called proxy noise by climate statisticians, owing to the many chemical bonds a carbon atom

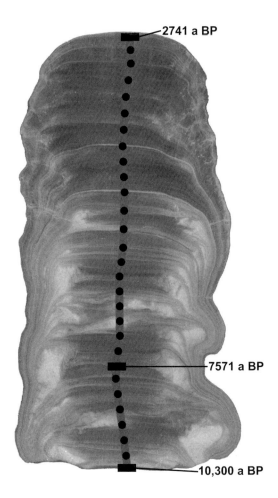

Figure B.1 Stalagmite (cross section). The growth direction is from bottom to top. Samples of typically a few mg are taken along the growth axis (thick gray line) using a dental drill; the majority of the samples (black circles) are used for measurement of the isotopic compositions ($\delta^{18}O$ and $\delta^{13}C$) by means of a mass spectrometer, some samples (black rectangles) are used for U/Th dating (Section B.5) by means of another mass spectrometer. The result is a proxy climate time series, such as shown in Figure 1.3. Speleothems are typically a few dm long. The sample symbols are not to scale. Adapted from Dominik Fleitmann (University of Reading, United Kingdom, personal communication, 2008). Reproduced with permission

can naturally assume. Cave monitoring (e.g., $\delta^{13}C$ in drip water) is one option to better understand proxy signals obtained from speleothems. Additional help comes from mathematically modeling the systems (Appendix C). This has been done to varying degrees: from modeling the global climate

(Werner et al. 2011) to modeling the carbon isotopes of carbonates in caves (Fohlmeister et al. 2011).

B.2 Lake Sediment Cores

Over time, the floor of a lake accumulates over time the solid material that is brought into the lake or is produced in it. The sediment carries time information (old bottom, young top), which can be retrieved by drilling a core. Proxy variables can be measured on core samples, which indicate past climates. Occasionally, the annual cycle is imprinted on the sediment as banded dark–bright layers, which are called varves (Figure B.2). For the sediment core from Lower Mystic Lake, Besonen et al. (2008) employed varve thickness as a proxy for hurricane occurrence (Section 2.4.1). Counting varves is a straightforward dating method, which is often augmented by dating results from radiometry to obtain a more robust chronology (Section B.5).

B.3 Tree-Rings

The time information in trees accumulates radially, and counting the annual tree-rings is one of the oldest dating tools in paleoclimatology. The reconstruction of a dated, long series from this archive is a challenge owing to the limited span a tree lives, which has led dendroclimatologists to utilize the data from several trees with overlapping spans (Figure B.3). By means of the overlap technique it is possible to go back in time; the current boundary is at about 14 ka BP (Reimer et al. 2013).

B.4 Other Natural Archives

In general, the accumulation process in a natural climate archive imprints the time information on the archive's material. Subsequently, paleoclimatologists and geoscientists sample the material, date it, and use it for proxy measurements. Natural archives may be formed by organisms, and there are more biological archives than tree-rings (Section B.3). Natural archives may also be of a geological or a glaciological type, and there are more geological archives than speleothems (Section B.1). We will discuss a few of these other archives: corals, marine sediment cores, and ice cores.

Hermatypic corals, which build massive carbonate reefs, are an important biological archive of information on sea-surface temperature (SST). This is because the coral polyp lives symbiotically with algae in the upper meters of

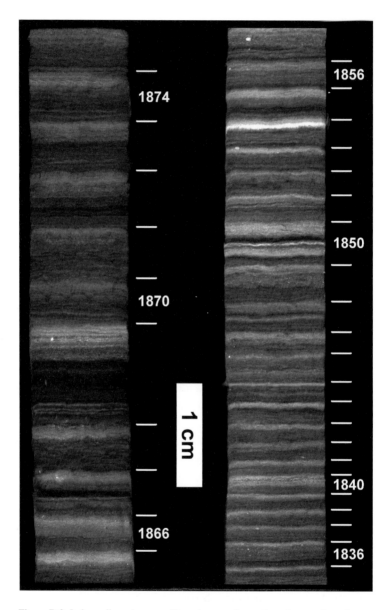

Figure B.2 Lake sediment varves. X-ray image of a section of a core from Lower Mystic Lake (Figure 2.6). Dark color means a larger proportion of biogenic sediment and bright color a larger proportion of siliciclastic material. The later varves are thicker owing to eutrophication driven by increased human population. From Besonen (2006). Reproduced with permission

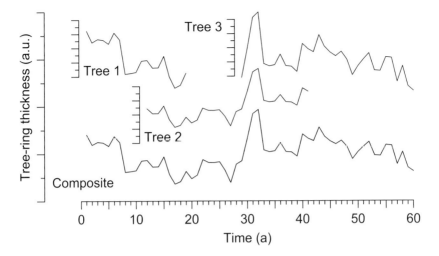

Figure B.3 Construction of a composite tree-ring record. This hypothetical example shows the ring-thickness records from three individual trees, which grew at overlapping periods. The ring-thickness variations reflect local effects (biology) and regional influences (climate); in the overlapping periods, the trees experienced similar conditions and, hence, show similar ring-thickness patterns. This allows the construction of a long composite record as the average of the standardized individual records. Standardization means subtraction of the sample mean and division by the sample standard deviation (Section D.1.1). The absolute thickness values are not of interest ("a.u." means "arbitrary units").

the ocean, and the production of carbonate depends on water temperature. The temperature information can be retrieved by measuring various proxy variables. The time information comes from U/Th dating (Section B.5) of the material, and occasionally from counting the coral's growth layers. Owing to the limited temperature range in which corals can live, the proxy SST information is geographically restricted to between about 30° north and south of the equator. Confounding variables may be water turbidity and cloudiness (Bradley 1999).

As an example for the use of corals, Felis et al. (2014) report the drilling into coral colonies at three sites on the Great Barrier Reef. The corals grew between the last glacial and the subsequent deglaciation, around 12 to 21 ka BP. The aim of the study was to assess the role of SST on the growth of the Great Barrier Reef during that time period. On the basis of U/Th dating and the SST proxies δ^{18}O and Sr/Ca ratio, it was found that the Great Barrier Reef developed through considerable SST change and "may be more resilient than previously thought" (Felis et al. 2014).

Marine sediments provide another geological archive, which is accumulated mainly by remainders of marine microbiota (e.g., shells of foraminifera) and terrigenous material brought in from rivers. A core drilled into the sediment retrieves the time information: old bottom, young top. To some degree this is distorted by the burrowing activity (bioturbation) of worms in the upper centimeters of the seafloor. The accumulation rate, the ratio of terrigenous to marine material, and the bioturbation depth vary geographically (Seibold and Berger 1996). Marine sediment cores are dated in several steps. First, a rough time frame is obtained by identifying core depths at which changes occurred. These comprise first or last appearance dates of certain microbiological species (e.g., foraminifera) and changes in the magnetic properties of sediment grains. Since the absolute dates of such changes are known (with some accuracy) from other sources, for example, compilations of reversals of the Earth's magnetic field (Gradstein et al. 2004), those sediment core depths constitute a frame of absolutely dated fixpoints (with some dating uncertainty). In a further step, the timescale between the fixpoints may be obtained by means of age–depth curve construction (Section B.5). The collection of marine sediment cores is a huge technical effort, which has been coordinated in various international research programs; see, for example, Ocean Drilling Program (1988–2007).

Snow, firn, and ice reflect the varying degrees of compaction and accumulation in Antarctica, the Arctic region, and elevated mountains. The time direction is preserved in the ice-core archive (old bottom, young top), but the ice layers are further compressed and thinned toward the bottom. In addition to that, the ice flows in dependence on the shape of the underlying bedrock. Compression and flow make construction of a timescale for drilled ice cores a challenge (Section B.5). Another obstacle for paleoclimatologists is the limited time interval accessible by means of ice cores. The currently oldest ice core (called EPICA) comes from Antarctica and goes back 800 ka. It is difficult to find older ice because of compaction, melting, and flow on the bedrock.

However, ice cores offer stunning proxy insights into past climates. This is owing to rather high time resolutions, which can be achieved for high-accumulation sites and with advanced laboratory techniques, such as continuous flow analysis, where the core is continuously melted along its axis and the proxy variable measured on the melt water. Ice cores further allow proxy information about the atmospheric composition of the past via enclosed air bubbles; however, those series have a considerably lower time resolution. One climate secret unravelled with the help of ice cores are the rapid warmings (called Dansgaard–Oeschger events) and subsequent coolings detected in high-resolution series from Greenland (Dansgaard et al. 1993).

> **Box B.1   Personal Reflection: Scientific jargon**
>
> The average spacing of a time series is $\bar{d} = [t(n) - t(1)]/(n-1)$. A series with a low $\bar{d}$ is called "highly resolved" by climatologists. Puzzled? I was so in the beginning, after I had moved from physics to geology for my PhD studies. But not very long. I learned about "ultra-high resolution" data, that geologists tend to be more enthusiastic (less self-critical?) about their work than physicists and that paleoclimatologists prefer to plot a climate record with nicely mild, warm conditions upward on the Y-axis (as in Figure 1.3). Do not get annoyed by jargon but immerse yourself in it. Do, however, take from time to time a sound distance and try to look at your discipline as an outsider, as someone from a neighboring discipline, or a philosopher of science. This helps to communicate your research to others – without invoking too much jargon.

B.5 Dating of Natural Archives

In the case a natural archive accumulated as layers over time, counting the layers is a straightforward dating tool. Annually varved archives have the desirable property that the spacing (one year) is exactly known. It is important to establish a geological–physical explanation for varve formation in an archive in order to achieve a sound chronology. Along these lines, the character of the lake sediment varves (Figure B.2) as an imprint of the annual change between biogenic sedimentation (growing season) and siliciclastic sedimentation (off season) was corroborated by Besonen (2006). If the start or the end date of the varve chronology is unknown ("floating"), absolute dating tools help.

Also ice cores may show annual cycles in certain variables, such as the NGRIP ice core in NH_4 (ammonium) and calcium (Wheatley et al. 2012). In fact, from the viewpoint of a paleoclimatologist, some variables may be measured more for dating purposes (e.g., NH_4), while other variables (e.g., $\delta^{18}O$) are relevant for climate inferences. Counting is often done by humans and – errare humanum est – this means counting errors. Layer recognition software (Wheatley et al. 2012) – itself not without uncertainties – is a help.

Absolute dating methods almost exclusively employ one of the many clocks provided by natural radioactive elements. Radiocarbon dating utilizes the decay of ^{14}C to ^{14}N, which has a half-life, $T_{1/2}$, of approximately 5730 a. The radiocarbon is formed in the atmosphere by bombarding cosmic-ray particles; soon it reacts to become part of atmospheric $^{14}CO_2$. Plants take $^{14}CO_2$ and

$^{12}CO_2$ up during photosynthesis to build their material – for example, a tree builds another ring. After the plant's system has been closed, the incorporated radiocarbon content is stored – for example, when a new tree-ring has been formed. Over time, the ^{14}C atoms in a tree-ring decay. The older a sample, the fewer ^{14}C atoms remain. Radiocarbon dating means (1) measuring the number of ^{14}C atoms (by means of decay counting or mass spectrometry) and (2) using the ratio of ^{14}C to ^{12}C at the time the material is formed. Determination of this initial $^{14}C/^{12}C$ ratio is often the harder challenge owing to the various forcings on radiocarbon: cosmic rays, Earth's magnetic field, and ocean overturning (Reimer et al. 2013).

That challenge of knowing the initial ratio applies also to U/Th dating, which utilizes the decays of ^{234}U to ^{230}Th ($T_{1/2} \approx 245$ ka) and ^{230}Th to ^{226}Ra ($T_{1/2} \approx 76$ ka). Here the speleothem archive offers an advantage since it contains essentially no thorium at the time of formation.

What age ranges can be realistically determined by means of radioactive dating? This is determined by the half-life. A too small age (compared with $T_{1/2}$) means nearly no decay; hence, it is difficult to quantify the deviation of the sample's radioactivity from the initial activity. A too large age means nearly no remaining radioactive atoms in a sample: hence, this is difficult to measure. A rule of thumb is to be cautious as regards ages smaller than about $0.1 \cdot T_{1/2}$ and larger than about $10 \cdot T_{1/2}$ (Mudelsee 2014). For the ages in between, the error in measuring the radioactive particles – and in turn the dating error – seems to be acceptably small.

How should you proceed if the archive does not show countable layers and instead a number of fixpoints, which have been absolutely dated with error bars? We wish to know the time values, $t(i)$, for all depths, $z(i)$, for which we have measurements, $x(i)$. This situation is rather common in paleoclimatology (Figure B.1).

Let us assume a number, n_{date}, of dated fixpoints, $\{z_{date}(j), t_{date}(j)\}_{j=1}^{n_{date}}$. It is possible to fit a number of growth curves through the dating points (Figure B.4). A linear curve type (Figure B.4b), obtained by means of linear regression (Section D.1.2), does not fit well; the two-piece linear model (Figure B.4c) is better. Geological–physical background knowledge about the archive is an important guide in selecting the type of growth curve. For example, speleothems may cease to grow for a period, and the two-piece linear model is a candidate; however, the model selection would have a more solid basis if also the double-checked sample revealed a change at around that depth point (e.g., by means of microscopy). If the background knowledge indicates a smooth model, not necessarily linear, a spline method (Figure B.4d) may be considered. There is one major constraint to growth-model selection: it has to be a monotonically increasing curve.

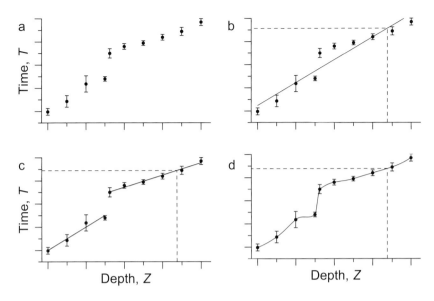

Figure B.4 Growth curves in a climate archive. The dated fixpoints, $\{z_{date}(j), t_{date}(j)\}_{j=1}^{n_{date}}$, are shown in each panel (filled symbols) with the standard errors (vertical lines); given are also the measurement points against depth, $\{z(i), x(i)\}_{i=1}^{n}$ (not shown). The task in this situation (a) is to infer $\{t(i)\}_{i=1}^{n}$. This is achieved by fitting a curve through the points, which describes the archive's growth. Shown (solid lines) are three curve types; (b) linear, (c) two-piece linear, and (d) Akima subspline interpolation. The inference (indicated by dashed lines) from $z(i)$ on $t(i)$ is done via the growth curve.

Reading Material

Felis et al. (2014) use corals as archive of the paleoclimate in the region of the Great Barrier Reef. Dansgaard et al. (1993) detected the Dansgaard–Oeschger events in the Greenland ice core GRIP. Röthlisberger et al. (2000) is a technical paper on continuous-flow analysis of ice cores. Wheatley et al. (2012) present an automated layer-counting method for dating ice cores. Besonen (2006) is a thesis, which contains detailed background information on the lake sediment core studied in the present book (Figure 2.6). Ocean Drilling Program (1988–2007) is a book series published by one of several research programs that have contributed a huge collection of marine sedimentary archives; see www-odp.tamu.edu (6 April 2019). Seibold and Berger (1996) is a textbook on oceanography and marine geology, with a focus on sediments and paleoclimate. Fairchild and Baker (2012) is a textbook on speleothems and their usage as archive in climatology. Reimer et al. (2013) present a reconstruction of past atmospheric radiocarbon from tree-rings, marine sediment cores, speleothems, and other climate archives.

Gradstein et al. (2004) is the standard chronology of geological and biological events (e.g., magnetic reversals, first and last appearance dates of species) on Earth covering the whole time range. Ivanovich and Harmon (1992) is the reference book on U/Th dating. Taylor (1987) is an accessible textbook on radiocarbon dating.

Mudelsee (2014: sections 4.1.7, 4.4, 9.1, and 9.6 therein) presents methods and reviews the literature on dating errors, growth curves, and timescale models. Engeln-Müllges and Reutter (1993) is a book on numerical algorithms with Fortran programs; its Appendix P 13.1 lists the code on Akima subspline interpolation, which is utilized for Figure B.4d.

Appendix C Physical Climate Models

Pierre Simon de Laplace (1749–1827) had the idea that, once we know the positions and momentums of all mass particles that constitute the world, we can predict its future. This mechanistic understanding in terms of first physical principles (position and momentum), combined with other first principles (conservation of energy and mass) is the principal basis for physical climate models.

The physical laws can be expressed as mathematical equations. How to solve these equations?

A straightforward approach to the equations leads to problems. Alone the atmosphere (simplified as ideal gas under standard pressure and temperature) has about 2.7×10^{19} particles per cm^3 (Loschmidt number). There is a knowledge problem (positions and momentums), and there is a computing problem. In addition, the Laplacian approach fails at the atomic level because of quantum mechanics (the uncertainty relation of position and momentum).

The key is to discretize the climate system spatially into grid boxes and to solve the equations for the grid averages; also temporally a discretization into time steps is employed. Resolvability is traded for computational feasibility. Still, it is possible to construct meaningful climate models for studying climate extremes (Section C.1), although the climate model output is inevitably noisy (Section C.2).

C.1 Earth System Models

Earth System Models (ESMs) are state-of-the-art climate models. They developed from Atmosphere–Ocean General Circulation Models (AOGCMs) to include various biogeochemical processes, such as the carbon cycle (Flato et al. 2013). That means, ESMs can calculate atmospheric CO_2 concentrations and account for the fluxes of this GHG between atmosphere, ocean, cryosphere, and biosphere (Eyring et al. 2016). Let us consider an atmospheric grid box in a typical ESM, which has a size in the order of 100 km (latitude) × 100 km (longitude) × 1 km (height). This single box contains about $2.7 \times 10^{19} \times 10^4$ km^3 cm^{-3} $= 2.7 \times 10^{38}$ particles. This degree of discretization (ten to the power of 38), and a time step in the order of one hour, achieve a feasible computational solution of the mathematical equations deemed necessary to describe the climate physics.

The discretization, however, means that some processes relevant for climate cannot be resolved by the model. For example, typical clouds are smaller than 100 km. Empirical formulas have to be plugged in to describe such sub-scale processes. To continue the example, the formation of clouds depends on humidity and temperature (grid-box averages). These empirical formulas are called parameterizations. They are obtained from separate experiments (measurements and modeling). The experiments exhibit measurement and modeling errors. Hence, the parameterizations are only approximate formulas. We call this error source parameterization uncertainty.

The Coupled Model Intercomparison Project Phase 6 (CMIP6) serves, among others, the grand challenge of "assessing climate extremes, what controls them, how they have changed in the past and how they might change in the future" (Eyring et al. 2016: p. 1944 therein).

Box C.1   **Personal Reflection: Laplace and the arrogance of physics**

As a physicist, I wholeheartedly agree with you that it appears arrogant to describe the climate system, if not the whole world, in a Laplacian manner as comprising point masses and nothing more. However, it was this type of reductionism that allowed physics to proceed by ignoring the disturbing noise and to detect the signal (the physical laws). Of course, the biological, geological, and chemical languages are, in many instances, better suited (i.e., they allow a more elegant, shorter, rational description) than physics – and they do find entrance into physical climate models via the parameterizations. If one models the climate, it is paramount to take into account the full suite of error sources and to report a result with an uncertainty measure (Section C.2).

We are perhaps satisfied to see that the blow to the Laplacian dream comes from two areas in physics. First, quantum mechanics tells us that on the atomic scale, position and momentum are not exactly defined (uncertainty relation). Second, the field of nonlinear dynamical systems informs us that any computer implementation of a model of the complex climate system, with its nonlinearly interacting variables (Figure 1.1), shows chaotic behavior – exponentially diverging solutions of the mathematical equations in dependence on the initial state. This uncertainty source, to be reported, is called internal variability.

C.2 Uncertainties and Model–Data Comparisons

An ESM tries to replicate the climate reality. It is a state-of-the-art, but not a perfect tool. The model output deviates from the climate reality. We can use the statistical language (Appendix D) to measure the size (standard error) and direction (bias) of the deviation. There are two uncertainty sources: internal variability (Box C.1) and parameterization (Section C.1). If an ESM is used for predicting future climate, then a third error source adds: the uncertainty about the radiative forcing (GHG emissions). If an ESM is used for projecting future climate, then the forcing uncertainty is ignored.

CMIP6, and other climate modeling initiatives, compare data and models in terms of standard errors and biases to assess the accuracy of climate predictions or projections. Determined biases help to improve the design of future models.

Mudelsee (2014: section 9.4.4 therein) considers a simple additive climate model bias. This allows a straightforward bias correction of the model output for the purposes of prediction or projection. We envisage that more complex stochastic models of the bias of ESMs (Jun et al. 2008) will lead to improved bias corrections.

Reading Material

The Fifth Assessment Report of the IPCC (Flato et al. 2013) gives a comprehensive description of climate models. The review by Giorgi and Gao (2018) considers future directions in the development of regional ESMs and identifies the inclusion of humans and the two-way interaction between society and nature as the next challenge. The journal Geoscientific Model Development has a special issue on CMIP6 (www.geosci-model-dev.net/special_issue590.html, 29 June 2019), where the current generation of ESMs participates.

Appendix D   Statistical Inference

We use statistical tools to make inferences about the climate reality. Estimation refers to a parameter in the climate equation (Eq. 1.2) that we wish to know about. Hypothesis testing is on whether a statement about the climate is true or not. Both actions of statistical inference – estimation (Section D.1) and hypothesis testing (Section D.2) – are associated with uncertainties. It is paramount for a climate researcher to determine the uncertainties and report them in the form of an error measure together with the result. Estimates without error bars are useless.

D.1 Estimation

Let θ be the parameter of interest in the climate equation for the process $\{X(i)\}$, and let $\widehat{\theta}$ be the estimator. The extension to a set of parameters is straightforward. Any meaningful construction lets the estimator be a function of the process, $\widehat{\theta} = g(\{X(i)\})$. (On the sample level, this means that it is wise to consider the data, $x(i)$, for estimation.) That means, $\widehat{\theta}$ is a random variable with statistical properties. The standard error is the standard deviation of $\widehat{\theta}$,

$$\text{se}_{\widehat{\theta}} = \left[VAR\left(\widehat{\theta}\right)\right]^{1/2}. \tag{D.1}$$

See Exercise 2.4 for definitions of expectation, variance, and standard deviation.

The bias of $\widehat{\theta}$ is

$$\text{bias}_{\widehat{\theta}} = E\left(\widehat{\theta}\right) - \theta. \tag{D.2}$$

$\text{bias}_{\widehat{\theta}} > 0$ ($\text{bias}_{\widehat{\theta}} < 0$) means a systematic overestimation (underestimation). $\text{se}_{\widehat{\theta}}$ and $\text{bias}_{\widehat{\theta}}$ are illustrated in Figure D.1. Desirable estimators have small

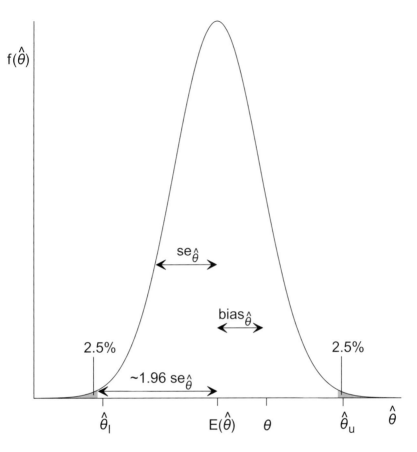

Figure D.1 Statistical properties of an estimator: standard error ($\mathrm{se}_{\widehat{\theta}}$), bias ($\mathrm{bias}_{\widehat{\theta}}$), and confidence interval ($\mathrm{CI}_{\widehat{\theta},\,1-2\alpha} = [\widehat{\theta}_\mathrm{l};\widehat{\theta}_\mathrm{u}]$) for a Gaussian distributed estimator, $\widehat{\theta}$. The true parameter value is θ; the confidence level is $1 - 2\alpha = 95\%$.

$\mathrm{se}_{\widehat{\theta}}$ and small $\mathrm{bias}_{\widehat{\theta}}$. In many estimations, a trade-off problem between $\mathrm{se}_{\widehat{\theta}}$ and $\mathrm{bias}_{\widehat{\theta}}$ occurs. A convenient measure is the root mean squared error,

$$
\begin{aligned}
\mathrm{RMSE}_{\widehat{\theta}} &= \left\{ E\left[\left(\widehat{\theta} - \theta\right)^2 \right] \right\}^{1/2} \\
&= \left(\mathrm{se}_{\widehat{\theta}}^2 + \mathrm{bias}_{\widehat{\theta}}^2 \right)^{1/2}.
\end{aligned}
\tag{D.3}
$$

The coefficient of variation is

$$
\mathrm{CV}_{\widehat{\theta}} = \mathrm{se}_{\widehat{\theta}} \big/ \left| E\left(\widehat{\theta}\right) \right|.
\tag{D.4}
$$

While $\widehat{\theta}$ is a best guess of θ or a point estimate, a confidence interval (CI) is an interval estimate that informs how good a guess is (Figure D.1). The CI for θ is

$$\mathrm{CI}_{\widehat{\theta},\,1-2\alpha} = \left[\widehat{\theta}_{\mathrm{l}}; \widehat{\theta}_{\mathrm{u}}\right], \tag{D.5}$$

where $0 \leq 1 - 2\alpha \leq 1$ is a prescribed value, denoted as confidence level. The practical examples in this book consider 90% ($\alpha = 0.05$) or 95% ($\alpha = 0.025$) CIs, which are reasonable choices for climatological problems. $\widehat{\theta}_{\mathrm{l}}$ is the lower, $\widehat{\theta}_{\mathrm{u}}$ the upper endpoint of the CI. $\widehat{\theta}_{\mathrm{l}}$ and $\widehat{\theta}_{\mathrm{u}}$ are random variables and have statistical properties such as standard error or bias.

The properties of interest for CIs are the coverages,

$$\gamma_{\mathrm{l}} = \mathrm{prob}\left(\theta \leq \widehat{\theta}_{\mathrm{l}}\right), \tag{D.6}$$

$$\gamma_{\mathrm{u}} = \mathrm{prob}\left(\theta \geq \widehat{\theta}_{\mathrm{u}}\right), \tag{D.7}$$

and

$$\gamma = \mathrm{prob}\left(\widehat{\theta}_{\mathrm{l}} < \theta < \widehat{\theta}_{\mathrm{u}}\right) = 1 - \gamma_{\mathrm{l}} - \gamma_{\mathrm{u}}. \tag{D.8}$$

Exact CIs have coverages, γ, equal to the nominal value, $1 - 2\alpha$. Construction of exact CIs requires knowledge of the distribution of $\widehat{\theta}$, which can be achieved only for simple problems, such as mean estimation of Gaussian white noise (Section D.1.1).

In more complex situations (i.e., for complex estimators, non-Gaussian distributions, and autocorrelation), approximate CIs have to be constructed by means of computational resampling algorithms such as the bootstrap (Mudelsee 2014). As regards the division of the nominal coverage between the CI endpoints, we adopt a practical approach and consider only equi-tailed CIs, where nominally $\gamma_{\mathrm{l}} = \gamma_{\mathrm{u}} = \alpha$. The second CI property besides coverage is interval length, $\widehat{\theta}_{\mathrm{u}} - \widehat{\theta}_{\mathrm{l}}$, which ideally is small.

Preceding paragraphs considered estimators on the process level. In practice, on the sample level, we plug in the data $\{t(i), x(i)\}_{i=1}^{n}$ for $\{T(i), X(i)\}_{i=1}^{n}$. Following the usual convention, we denote also the estimator on the sample level as $\widehat{\theta}$.

D.1.1 Mean and Standard Deviation, Standardization

Let the process $\{X(i)\}_{i=1}^{n}$ be given by

$$X(i) = \mathcal{E}_{\mathrm{N}(\mu,\,\sigma^2)}(i), \qquad i = 1, \ldots, n, \tag{D.9}$$

which is called a Gaussian purely random process or Gaussian white noise. There is no serial dependence, and the times, $T(i)$, are not of interest. This process is called white noise because its spectrum (power against frequency) is flat, without a preference for slow or fast variations (Mudelsee 2014: chapter 5 therein); white light has a flat color spectrum. Consider the sample mean, $\widehat{\mu}$, as estimator of the mean, μ. On the process level, we write

$$\widehat{\mu} = \bar{X} = \sum_{i=1}^{n} X(i)/n. \tag{D.10}$$

Let also σ be unknown and estimated by the sample standard deviation, $\widehat{\sigma} = S_{n-1}$, given in Eq. (D.19). The properties of $\bar{X}$ readily follow as

$$\text{se}_{\bar{X}} = \sigma \cdot n^{-1/2}, \tag{D.11}$$

$$\text{bias}_{\bar{X}} = 0, \tag{D.12}$$

$$\text{RMSE}_{\bar{X}} = \text{se}_{\bar{X}}, \tag{D.13}$$

and

$$\text{CV}_{\bar{X}} = \sigma \cdot n^{-1/2} \cdot \mu^{-1}. \tag{D.14}$$

An exact CI of level $1 - 2\alpha$ can be constructed by means of the Student's t distribution of $\bar{X}$,

$$\text{CI}_{\bar{X}, 1-2\alpha} = \left[\bar{X} + t_{n-1}(\alpha) \cdot S_{n-1} \cdot n^{-1/2}; \bar{X} + t_{n-1}(1-\alpha) \cdot S_{n-1} \cdot n^{-1/2} \right]. \tag{D.15}$$

$t_{\nu}(\beta)$ is the percentage point at β of the t distribution function with ν degrees of freedom, which can be found in textbooks (Mudelsee 2014).

On the sample level, we write the estimated sample mean,

$$\widehat{\mu} = \bar{x} = \sum_{i=1}^{n} x(i)/n, \tag{D.16}$$

the estimated standard error,

$$\widehat{\text{se}}_{\bar{x}} = \left\{ \sum_{i=1}^{n} [x(i) - \bar{x}]^2 / n^2 \right\}^{1/2}, \tag{D.17}$$

and the confidence interval,

$$\text{CI}_{\bar{x}, 1-2\alpha} = \left[\bar{x} + t_{n-1}(\alpha) \cdot s_{n-1} \cdot n^{-1/2}; \bar{x} + t_{n-1}(1-\alpha) \cdot s_{n-1} \cdot n^{-1/2} \right], \tag{D.18}$$

where s_{n-1} is given by Eq. (D.25).

Next, consider as estimator of σ the sample standard deviation with denominator $n - 1$, written on the process level as

$$\widehat{\sigma} = S_{n-1} = \left\{ \sum_{i=1}^{n} \left[X(i) - \bar{X} \right]^2 / (n - 1) \right\}^{1/2}. \qquad \text{(D.19)}$$

The properties of S_{n-1} are as follows:

$$\mathrm{se}_{S_{n-1}} = \sigma \cdot \left(1 - c^2 \right)^{1/2}, \qquad \text{(D.20)}$$

$$\mathrm{bias}_{S_{n-1}} = \sigma \cdot (c - 1), \qquad \text{(D.21)}$$

$$\mathrm{RMSE}_{S_{n-1}} = \sigma \cdot [2(1 - c)]^{1/2}, \qquad \text{(D.22)}$$

and

$$\mathrm{CV}_{S_{n-1}} = \left(1/c^2 - 1 \right)^{1/2}, \qquad \text{(D.23)}$$

where

$$c = [2/(n-1)]^{1/2} \cdot \Gamma(n/2) \Big/ \Gamma((n-1)/2) \qquad \text{(D.24)}$$

and Γ is the gamma function (Mudelsee 2014).

On the sample level, we write

$$\widehat{\sigma} = s_{n-1} = \left\{ \sum_{i=1}^{n} [x(i) - \bar{x}]^2 / (n - 1) \right\}^{1/2} \qquad \text{(D.25)}$$

and use the chi-squared distribution of S_{n-1}^2 to find

$$\mathrm{CI}_{s_{n-1},\,1-2\alpha} = \left[s_{n-1} \left[(n-1) \Big/ \chi_{n-1}^2(\alpha) \right]^{1/2} ; \right.$$

$$\left. s_{n-1} \left[(n-1) \Big/ \chi_{n-1}^2(1-\alpha) \right]^{1/2} \right], \qquad \text{(D.26)}$$

where $\chi_\nu^2(\beta)$ is the percentage point at β of the chi-squared distribution function with ν degrees of freedom (Mudelsee 2014).

The standardization of a sample, $\{x(i)\}_{i=1}^{n}$, means the calculation of the expression $\{[x(i) - \bar{x}]/s_{n-1}\}_{i=1}^{n}$. This is often done for comparing series with different location and spread, such as in Figure B.3.

D.1.2 Linear Regression

Consider the climate equation (Eq. 1.2) with a linear trend component, absent extreme component, and constant variability,

$$X(i) = \beta_0 + \beta_1 T(i) + S \cdot X_{\text{noise}}(i). \tag{D.27}$$

T is called the predictor or regressor variable, X the response variable, and β_0 and β_1 the regression parameters.

The task is to estimate β_0 and β_1 using the sample, $\{t(i), x(i)\}_{i=1}^{n}$, that means, to find the straight line that is closest to the data. One option to measure distance is via the sum of the squares of the distances between the line and a data point, $x(i)$. (Taking the squares means it does not matter if the line is above or below the data point.) The ordinary least-squares (OLS) estimation, hence, minimizes

$$SSQ(\beta_0, \beta_1) = \sum_{i=1}^{n} [x(i) - \beta_0 - \beta_1 t(i)]^2. \tag{D.28}$$

Setting the first derivatives of $SSQ(\beta_0, \beta_1)$ with respect to β_0 and β_1 equal to zero and solving the two resulting equations for the two unknowns yields the OLS estimators,

$$\widehat{\beta_0} = \left[\sum_{i=1}^{n} x(i) - \widehat{\beta_1} \sum_{i=1}^{n} t(i) \right] \bigg/ n, \tag{D.29}$$

$$\widehat{\beta_1} = \left[\sum_{i=1}^{n} t(i) \sum_{i=1}^{n} x(i) \bigg/ n - \sum_{i=1}^{n} t(i) x(i) \right]$$

$$\times \left[\left(\sum_{i=1}^{n} t(i) \right)^2 \bigg/ n - \sum_{i=1}^{n} t(i)^2 \right]^{-1}. \tag{D.30}$$

If the linear model is suitable and the noise component in the regression equation (Eq. D.27) is Gaussian white noise, then the standard errors of the regression parameters can be analytically determined,

$$se_{\widehat{\beta_0}} = MS_E^{1/2} \left\{ 1/n + \left(\sum_{i=1}^{n} t(i)/n \right)^2 \bigg/ \left[\sum_{i=1}^{n} t(i)^2 - \left(\sum_{i=1}^{n} t(i) \right)^2 \bigg/ n \right] \right\}^{1/2}, \tag{D.31}$$

$$se_{\widehat{\beta_1}} = MS_E^{1/2} \left[\sum_{i=1}^{n} t(i)^2 - \left(\sum_{i=1}^{n} t(i) \right)^2 \bigg/ n \right]^{-1/2}, \tag{D.32}$$

where

$$MS_E^{1/2} = \left\{ \sum_{i=1}^{n} \left[x(i) - \widehat{\beta}_0 - \widehat{\beta}_1 t(i) \right]^2 \Big/ (n-2) \right\}^{1/2}.$$ (D.33)

MS_E is called residual mean square, and $MS_E^{1/2}$ can be used as an estimator of the variability, S.

If the linear model is suitable and the noise component in the regression equation is complex – regarding distributional shape or serial dependence – it may be that the task of determining the standard errors of the regression parameters becomes analytically intractable. Resampling methods (Mudelsee 2014) help in such a situation.

D.1.3 Nonlinear Regression

Consider the climate equation (Eq. 1.2) with a nonlinear trend component (Figure D.2), absent extreme component, and time-dependent variability,

$$X(i) = X_{\text{ramp}}(i) + S(i) \cdot X_{\text{noise}}(i),$$ (D.34)

where

$$X_{\text{ramp}}(T) = \begin{cases} x1 & \text{for } T \le t1, \\ x1 + (T - t1)(x2 - x1)/(t2 - t1) & \text{for } t1 < T \le t2, \\ x2 & \text{for } T > t2, \end{cases}$$ (D.35)

and $X_{\text{ramp}}(i)$ is the discrete-time version of $X_{\text{ramp}}(T)$. The ramp is an example of a nonlinear regression model. It has four parameters: start time $t1$, start level $x1$, end time $t2$, and end level $x2$.

The task is to estimate the parameters using the sample. One estimation option is weighted least-squares (WLS), that means, the minimization of

$$SSQW(t1, x1, t2, x2) = \sum_{i=1}^{n} \left[x(i) - x_{\text{ramp}}(i) \right]^2 \Big/ S(i)^2,$$ (D.36)

where $x_{\text{ramp}}(i)$ is the sample version of $X_{\text{ramp}}(i)$.

There is a mathematical inconvenience associated with the ramp: the two bends at $t1$ and $t2$ (Figure D.2), where the curve is not differentiable. This means that the first derivatives of $SSQW$ with respect to $t1$ and $t2$ are not defined. Hence, there exist no analytical solutions for $\widehat{t1}$ and $\widehat{t2}$. The original paper on the ramp (Mudelsee 2000) choose a pragmatic, unelegant

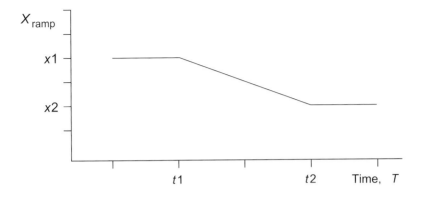

Figure D.2 Ramp regression model. It has four parameters: $t1$, $x1$, $t2$, and $x2$.

approach, namely trying all possible combinations of $t1 \in \{t(i)\}_{i=1}^{n}$, $t2 \in \{t(i)\}_{i=1}^{n}$ and taking the combination that minimizes $SSQW(t1, x1, t2, x2)$. (For fixed $t1$ and $t2$, there exist analytical solutions for $x1$ and $x2$.) This approach, which is called brute-force or exhaustive search (Appendix E), is feasible for data sizes up to, say, 1,000,000 and modern computers. The non-differentiability further requires bootstrap resampling (Mudelsee 2000) to determine uncertainty measures for the estimated ramp parameters.

Why the weighting, the division by $S(i)^2$ for the ramp fitting (Eq. D.36)? It can be shown for simpler problems, such as linear regression, that WLS yields smaller estimation standard errors than OLS (Montgomery and Peck 1992). In other words, it pays to put more weight on more accurate data points, $x(i)$, which have smaller $S(i)$. However, $S(i)$ has to be known or at least estimated.

Box D.1   **Personal Reflection: The sample standard deviation, $n-1$, and the freedom which estimator to select**

Eq. (D.19) presents one standard deviation estimator: S_{n-1}, that means, the sample standard deviation with denominator $n-1$. This estimator has certain statistical properties, of which the standard error, bias, RMSE, CV, and CI are given for the case the data generating process is Gaussian white noise.

However, there is another estimator: the sample standard deviation with denominator n,

$$\hat{\sigma} = S_n = \left\{ \sum_{i=1}^{n} \left[X(i) - \bar{X} \right]^2 / n \right\}^{1/2}. \qquad (D.37)$$

This estimator also has a standard error, bias, and so forth.

Which estimator to select, S_{n-1} or S_n? Of course we wish to select the estimator that gives results closer to the true but unknown value, σ. In general, the chosen estimator depends on the data generating process (e.g., distributional shape). My answer to the question is as follows. For practical applications it does not matter whether you take S_{n-1} or S_n because n is usually not too small and the resulting differences are likely to be much smaller than the effects of the violations of the distributional assumption.

I remember well the practical courses I took during my semesters as a student of physics. The chief course instructor was said to be very strict. We were doing repeated measurements. We were told that one has to use S_{n-1} and not S_n. This was nonsense. Perhaps it was motivated by the fact that the estimator, S_{n-1}^2, of the variance, σ^2, given by

$$S_{n-1}^2 = \sum_{i=1}^{n} \left[X(i) - \bar{X} \right]^2 / (n - 1), \qquad (D.38)$$

has bias$_{S_{n-1}^2} = 0$ for Gaussian white noise. Zero bias is certainly nice, but small RMSE is more important. Furthermore, note that bias$_{S_{n-1}^2} = 0$ does not mean that bias$_{S_{n-1}} = 0$, as Eq. (D.21) informs us. In retrospect, it would have been wiser if the instructors had told us that we could chose which estimator to select, and that in physics S_{n-1} is used for conventional reasons.

The essence of this story is that it is not prescribed which estimator to use. In the case of standard deviation estimation of Gaussian white noise this has a negligible effect. However, the life of a climate researcher is more interesting: there are more complex things to be estimated than the standard deviation, and there are more complex random components than Gaussian white noise. In this real climate world, it is often not at all clear which the best estimator is in terms of a certain measure (standard error, bias, RMSE, robustness, and so forth). Monte Carlo experiments (Section D.1.4), where the true values an estimator should guess are known, help to compare estimators.

A final word on estimators. How are they invented? My answer: intuitively. This has something to do with creativity. This is where the fun is for statisticians.

D.1.4 Monte Carlo Experiments

A Monte Carlo experiment is a computer simulation of a random process. The properties of the data generating process, such as mean and standard deviation, can be prescribed. Hence, unlike in the real world, the truth is known. The random component is brought in by means of a random number generator (Section E.3). Since the prescribed parameter values are known, Monte Carlo experiments can be used as objective tests of the estimators of the parameters and of CI construction methods.

Table D.1 shows the results of a Monte Carlo experiment on a simple task, mean estimation of Gaussian white noise. A number of $n_{\text{sim}} = 4,750,000$ random samples of $\{X(i)\}_{i=1}^{n}$ were generated according to Eq. (D.9) with $\mu = 1.0$, $\sigma = 2.0$, and various n values. A confidence interval, $\text{CI}_{\bar{x}, 1-2\alpha}$, was constructed for each simulation using Eq. (D.18) with $\alpha = 0.025$. Empirical RMSE (denoted as $\text{RMSE}_{\bar{x}}$) and empirical coverage ($\gamma_{\bar{x}}$) were determined subsequently as follows.

$\text{RMSE}_{\bar{x}}$ is given by $\left[\sum_{i=1}^{n_{\text{sim}}} (\bar{x} - \mu)^2 / n_{\text{sim}} \right]^{1/2}$. Its nominal value is (Eq. D.11) equal to $\sigma \cdot n^{-1/2}$.

$\gamma_{\bar{x}}$ is given by the number of simulations where $\text{CI}_{\bar{x}, 1-2\alpha}$ contains μ, divided by n_{sim}. The standard error of $\gamma_{\bar{x}}$ follows from the binomial distribution (Efron and Tibshirani 1993); it is nominally equal to $\left[2\alpha(1 - 2\alpha)/n_{\text{sim}} \right]^{1/2} = 0.0001$.

Table D.1. *Monte Carlo experiment, mean estimation of a Gaussian white noise process. The entries are rounded.*

n	$\text{RMSE}_{\bar{x}}$	Nominal	$\text{RMSE}_{\bar{x}_{\text{nonsense}}}$	Nominal	$\gamma_{\bar{x}}$	Nominal
10	0.6327	0.6325	0.7562	0.7559	0.9499	0.9500
20	0.4474	0.4472	0.5002	0.5000	0.9498	0.9500
50	0.2828	0.2828	0.3049	0.3050	0.9501	0.9500
100	0.2000	0.2000	0.2108	0.2108	0.9499	0.9500
200	0.1415	0.1414	0.1467	0.1466	0.9499	0.9500
500	0.0894	0.0894	0.0915	0.0915	0.9500	0.9500
1000	0.0633	0.0632	0.0643	0.0643	0.9499	0.9500

Also given is the empirical RMSE for another estimator, given on the sample level by

$$\bar{x}_{\text{nonsense}} = \sum_{i=1}^{n_{\text{nonsense}}} X(i)/n_{\text{nonsense}}, \qquad (\text{D.39})$$

where $n_{\text{nonsense}} = NINT(n - n^{1/2})$ and $NINT$ is the nearest integer function.

We learn two things from the Monte Carlo experiment (Table D.1). First, the estimator $\bar{X}$ performs exactly as expected, and also the CI for it is exact; the only differences between achieved and nominal values are due to simulation noise (n_{sim} is less than infinity). Second, it is nonsense to throw away $\sqrt{n}$ data points; this is punished by a larger RMSE (of $\bar{X}_{\text{nonsense}}$ compared with $\bar{X}$).

D.2 Hypothesis Tests

A hypothesis test, also called significance test or statistical test, involves the following procedure (Mudelsee 2014: section 3.6 therein). A null hypothesis (or short: null), H_0, is formulated. H_0 is tested against an alternative hypothesis, H_1. The hypotheses H_0 and H_1 are mutually exclusive. H_0 is a simple null hypothesis if it completely specifies the data generating process, for example, "$X(i)$ is Gaussian white noise with zero mean and unit standard deviation." H_0 is a composite null hypothesis if some parameter of $X(i)$ is unspecified, for example, "Gaussian white noise with zero mean." Next, a test statistic, U, is calculated. Any meaningful construction lets U be a function of the data generating process, $U = g\left(\{T(i), X(i)\}_{i=1}^{n}\right)$. On the sample level, $u = g\left(\{t(i), x(i)\}_{i=1}^{n}\right)$. (This means that it is wise to consider the data, $x(i)$, for hypothesis testing.) In case of the Cox–Lewis test (Section 3.2.5), we specify the notation and write U_{CL} and u_{CL}. In the example of H_0: "Gaussian white noise process with $\mu = 0$," one could take $U = \bar{X} = \sum_{i=1}^{n} X(i)/n$, the sample mean. U is a random variable with a distribution function, $F_0(u)$, where the index "0" indicates that U is computed "under H_0," that is, as if H_0 were true. $F_0(u)$ is the null distribution. In the example, $F_0(u)$ would be Student's t distribution function. If in the example the alternative were H_1: "$\mu > 0$," then a large, positive u value would speak against H_0 and for H_1. Using $F_0(u)$ and plugging in the data, $\{t(i), x(i)\}_{i=1}^{n}$, the one-sided significance probability or one-sided P-value results as

$$P = \text{prob}\,(U \geq u \mid H_0)$$
$$= 1 - F_0(u). \qquad (\text{D.40})$$

The P-value is the probability that under H_0 a value of the test statistic greater than or equal to the observed value, u, is observed. If P is small, then H_0 is rejected and H_1 accepted, otherwise H_0 cannot be rejected and H_1 cannot be accepted. The two-sided P-value is

$$P = \text{prob}\left(|U| \geq |u| \mid H_0\right). \tag{D.41}$$

In the example, a two-sided test would be indicated for H_1: "Gaussian white noise with $\mu \neq 0$." Besides the P-value, a second result of a statistical test is the power. In the one-sided test example:

$$\text{power} = \text{prob}\left(U \geq u \mid H_1\right). \tag{D.42}$$

A type-2 error is accepting H_0, although it is a false statement and H_1 is true. The probability of a type-2 error is $\beta = 1 - \text{power}$. A type-1 error is rejecting H_0 against H_1, although H_0 is true. P, the significance probability, is therefore denoted also as type-1-error probability or false-alarm probability; u is denoted also as false-alarm level.

Although H_0 can be a composite null, it is usually more explicit than H_1. In climatological practice, the selection of H_1 should be guided by prior climatological knowledge. H_1 determines also whether a test should be designed as one- or two-sided. For example, if H_0 were "no temperature change in a climate model experiment studying the effects of doubled CO_2 concentrations, $\Delta T = 0$," then a one-sided test against H_1: "$\Delta T > 0$" would be indicated because physics would not let one expect a temperature decrease. Because H_1 is normally rather general, it is difficult to quantify the test power. Therefore, more emphasis is put on accurate P-values. Various test statistics, $U_1, U_2, \ldots$, may be appropriate for testing H_0 against H_1. The statistic of choice has, for a given dataset, a small type-1-error probability (small P-value) as first quality criterion. The second quality criterion is a small type-2-error probability (large power), preferably calculated for some realistic, explicit alternative. The design of a test statistic is, for complex problems, something that has to do with creativity. We can say that a test does not intend to prove that a hypothesis is true but rather that it tries to reject a null hypothesis. A hypothesis becomes more reliable after it has been tested successfully against realistic alternatives.

Here are some critical remarks. It is important that H_0 and H_1 are established independently of the data to prevent circular reasoning (von Storch and Zwiers 1999: section 6.4 therein). It is more informative to give P-values than to report merely whether they are below certain specified significance levels, say $P < 0.1, 0.05$, or 0.01. If a hypothesis test is repeated several times on different data, it becomes more likely to find a significant single

result. Such multiple tests therefore require corrections to the employed
P-values (Mudelsee 2014: section 5.2.5 therein). Discarding "unsuccessful"
tests and reporting only the "significant" effects, which is sometimes called
"P-hacking," is a case of scientific misconduct – it does not warrant to
completely abandon hypothesis tests (as has been done by a psychology journal
in 2015). Hypothesis tests should rather be seen as a guide that helps us to
better assess estimation results. However, estimation is more important than
hypothesis testing because it tells us not merely whether there is an effect but
how large it is.

Reading Material

Brockwell and Davis (1991, 1996) are textbooks on time series analysis; the
second one is accessible to non-statisticians, the first one not immediately.
Chatfield (2004) is an accessible textbook on time series analysis, which
contains exercises and solutions. Efron and Tibshirani (1993) is a textbook
on bootstrap resampling and CI construction. Fishman (1996) is a textbook on
Monte Carlo experiments, which contains algorithms. Johnson et al. (1994)
and Johnson et al. (1995) are both parts of a series of reference books on
statistical distributions. Montgomery and Peck (1992) is a textbook on linear
regression. Mudelsee (2014) is a book on the analysis of climate time series by
means of bootstrap resampling and CI construction. It contains algorithms. The
present chapter on inference has (slightly modified) excerpts from Mudelsee
(2014: sections 3.1.1, 3.1.2, and 3.6 therein). The paper by Mudelsee (2000)
introduces ramp regression with bootstrap resampling to quantify climate
transitions. Finally, von Storch and Zwiers (1999) is a textbook on the analysis
of climate data.

Appendix E   Numerical Techniques

Numerical techniques help to calculate or generate numbers, usually by means of a computer. One situation is when we have data and wish to use an algorithm for the calculation of an estimate. In the real-world climate, the noise component (Eq. 1.2) is complex, it displays non-Gaussian distributions and persistence; the estimation procedure often is also complex (nonlinear functions, etc.); see Box D.1.

Hence, in this situation it is usually not possible to analytically write down the exact result. Under such analytical intractability we turn to a numerical procedure, which approaches the solution (i.e., the true numerical value) in small and many steps. A typical case of such a complex estimation occurs when a function that depends nonlinearly on several variables has to be minimized (e.g., least-squares estimation); this case is called optimization (Section E.1).

The computer hardware influences what numerical techniques are feasible. Since the physical speed limit of single number processing units has been reached, the trend in the past years has been to construct multiprocessor computers. Software, lagging behind hardware in development, has taken up the trend. Monte Carlo simulations (Section D.1.4) are particularly suited to multiprocessor machines because the many simulations can run parallel. This is explained in Section E.2. These stochastic simulations require random number generators (Section E.3).

E.1 Optimization

Consider the log-likelihood function, l, of the time-dependent GEV distribution (Eqs. 3.13–3.15),

$$l(\beta_0, \beta_1, \gamma_0, \gamma_1, \delta_0, \delta_1). \tag{E.1}$$

In the six-dimensional space that is spanned by the parameters, the log-likelihood function has a maximum at the point

$$\widehat{\boldsymbol{\theta}} = (\widehat{\beta}_0, \widehat{\beta}_1, \widehat{\gamma}_0, \widehat{\gamma}_1, \widehat{\delta}_0, \widehat{\delta}_1)'. \tag{E.2}$$

This point, $\widehat{\boldsymbol{\theta}}$, is the maximum likelihood estimate. (The prime denotes the transpose.) Usually, $\widehat{\boldsymbol{\theta}}$ cannot be analytically calculated. It has to be searched for by means of an optimization technique.

The gradient technique is one optimizer applicable here. It proceeds as follows.

Step 1 Make an initial guess, $\widehat{\boldsymbol{\theta}}^{(0)}$, where to start.

Step 2 Calculate the log-likelihood at that point, $l\left(\widehat{\boldsymbol{\theta}}^{(0)}\right)$.

Step 3 Calculate the gradient of the log-likelihood at that point, $\nabla\left[l\left(\widehat{\boldsymbol{\theta}}^{(0)}\right)\right]$.

Step 4 The gradient is a vector that points in the direction of the steepest increase of the log-likelihood function. (See Figure E.1 for an illustration.) Go a step of a certain size in that direction (the closer to the maximum, the smaller the step size). The new point is $\widehat{\boldsymbol{\theta}}^{(1)}$.

Step 5 Recalculate the log-likelihood function, $l\left(\widehat{\boldsymbol{\theta}}^{(1)}\right)$, and the gradient, $\nabla\left[l\left(\widehat{\boldsymbol{\theta}}^{(1)}\right)\right]$, at that point.

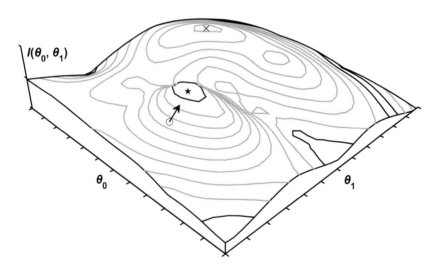

Figure E.1 Two-dimensional optimization problem. The log-likelihood function to be maximized, l, depends on two parameters, θ_0 and θ_1. Shown are the contour lines (gray or black) of $l(\theta_0, \theta_1)$; the overall maximum point (asterisk), $\widehat{\boldsymbol{\theta}}$; the current point (open circle), $\widehat{\boldsymbol{\theta}}^{(j)}$; and the gradient (arrow), $\nabla\left[l\left(\widehat{\boldsymbol{\theta}}^{(j)}\right)\right]$. The gradient optimization technique helps to approach the maximum. Also shown in this two-dimensional landscape is a local maximum (cross).

Step 6 Check the stopping rule. One possible formulation of a rule is that the log-likelihood has not changed much,

$$l\left(\widehat{\boldsymbol{\theta}}^{(j)}\right) - l\left(\widehat{\boldsymbol{\theta}}^{(j-1)}\right) \approx 0. \tag{E.3}$$

(j is a counter.) Another rule is that the absolute value of the gradient is close to zero,

$$\left|\nabla\left[l\left(\widehat{\boldsymbol{\theta}}^{(j)}\right)\right]\right| \approx 0. \tag{E.4}$$

"Close to zero" is determined by the accuracy of numbers in the computer; on 32-bit machines, the accuracy is in the order of eight digits. Thus, if the rule informs that a solution is sufficiently close to a maximum, then stop and take the approximate estimate, $\widehat{\boldsymbol{\theta}}^{(j)}$. If not, then increase the counter by one and go back to Step 2.

In case of the maximum likelihood estimation, the gradient technique advantageously produces also the approximate standard errors of the parameters. This goes via the curvature of the log-likelihood function (Coles 2001: section 2.6.4 therein). (While the gradient is calculated from the first derivatives of the log-likelihood function, the curvature is calculated from the second derivatives.) Note that those errors are approximate: assumed is a large data size, for which the estimator's distribution becomes normal.

The gradient technique is not applicable to each estimation problem. For example, the WLS sum for ramp estimation (Eq. D.36) is not differentiable with respect to two parameters ($t1$ and $t2$). The brute-force search for $\widehat{t1}$ and $\widehat{t2}$ (Section D.1.3) is feasible for typical data sizes in climate sciences and modern computers. The brute-force search advantageously produces a global optimum; that means, we can be sure that there are no better solutions. On the other hand, the gradient technique may land us in a local optimum – there may be another hill, $\widehat{\boldsymbol{\theta}}$, that is actually higher than the hill we are currently standing on, $\widehat{\boldsymbol{\theta}}^{(j)}$. The existence of such other solutions may be examined by employing a variety of starting points, $\widehat{\boldsymbol{\theta}}^{(0)}$, distributed over the landscape.

Optimization problems where brute force is not feasible require search heuristics for their solution. Genetic algorithms constitute an interesting approach. Let the current search state be composed of a population of k individual potential solutions,

$$\boldsymbol{P}^{(i)} = \left\{\widehat{\boldsymbol{\theta}}_1^{(i)}, \widehat{\boldsymbol{\theta}}_2^{(i)}, \ldots, \widehat{\boldsymbol{\theta}}_k^{(i)}\right\}. \tag{E.5}$$

Each solution corresponds to a measure (e.g., *SSQW*). This generation undergoes variation. For example, from two individual solutions ("parents") a new individual solution ("offspring") may be formed under some variation rule. The new generation is

$$P^{(i+1)} = \left\{ \widehat{\theta}_1^{(i+1)}, \widehat{\theta}_2^{(i+1)}, \ldots, \widehat{\theta}_k^{(i+1)} \right\}.$$ (E.6)

Some members of the new generation have been kept (e.g., $\widehat{\theta}_1^{(i+1)} = \widehat{\theta}_1^{(i)}$), while others have been replaced by a varied solution. The idea is to keep the good properties (in terms of the measure) of the current generation – not of just one solution but of a pool – and to use the pool to find better solutions. It is obvious that the implementation of search heuristics for optimization requires suitably selecting k, a measure, and variation rules (Michalewicz and Fogel 2000).

A note on the computer language in which to program optimization code. Fortran and C/C++ are low-level languages, closer to the representation of numbers by bits and bytes in the computer. Hence, they allow a better control over numerical accuracy – crucial for optimization problems – than high-level languages and packages, such as Excel, Matlab, or S-Plus/R.

E.2 Parallel Computing

A multiprocessor/multicore computer has many computing units, which can work independently of each other at the same time. How many? Today, typical tablets or laptops have a few processors, workstations a few tens, and supercomputers a few thousand. Parallel computing on such machines can make big calculations feasible. Note that a doubled number of processors does not mean that the run time is halved. It means a longer span because management of the parallel runs and evaluation of the summary results is necessary. Possibly, the independence is violated to some degree, and one processor may have to wait for another's result.

Monte Carlo simulations (Section D.1.4) are particularly suited because the many repetitions of generating data and calculating an estimate are completely independent of each other. The key requisite is that the streams of the random numbers on the processors are independent, which can be ensured by suitably seeding the random number generator (Figure E.2 and Section E.3).

A note on the software implementation of parallel computing. The low-level languages Fortran and C/C++ can still be used. The computer code needs to be augmented by language directives, which determine the number

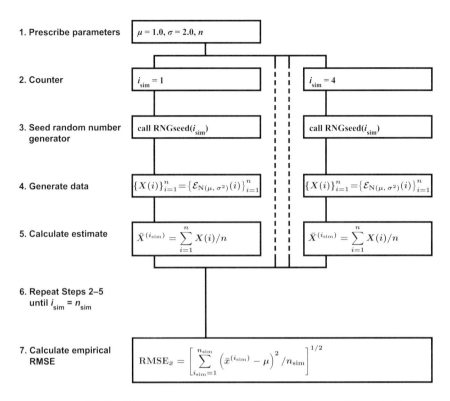

1. Prescribe parameters $\mu = 1.0, \sigma = 2.0, n$

2. Counter $i_{\text{sim}} = 1$ $i_{\text{sim}} = 4$

3. Seed random number generator call RNGseed(i_{sim}) call RNGseed(i_{sim})

4. Generate data $\{X(i)\}_{i=1}^{n} = \{\mathcal{E}_{N(\mu, \sigma^2)}(i)\}_{i=1}^{n}$ $\{X(i)\}_{i=1}^{n} = \{\mathcal{E}_{N(\mu, \sigma^2)}(i)\}_{i=1}^{n}$

5. Calculate estimate $\bar{X}^{(i_{\text{sim}})} = \sum_{i=1}^{n} X(i)/n$ $\bar{X}^{(i_{\text{sim}})} = \sum_{i=1}^{n} X(i)/n$

6. Repeat Steps 2–5 until $i_{\text{sim}} = n_{\text{sim}}$

7. Calculate empirical RMSE
$$\text{RMSE}_{\bar{x}} = \left[\sum_{i_{\text{sim}}=1}^{n_{\text{sim}}} \left(\bar{x}^{(i_{\text{sim}})} - \mu \right)^2 / n_{\text{sim}} \right]^{1/2}$$

Figure E.2 Parallel computing in a Monte Carlo experiment. Schematically shown from top to bottom is the flow of computation steps in the experiment on mean estimation of Gaussian white noise (Section D.1.1). The implementation is for a computer that allows for four threads of parallel computation (indicated by vertical solid/dashed lines). Step 3 contains a call to a seeding routine of the random number generator, which is necessary to ensure independence of data among simulations (Section E.3). Note that Steps 4 and 5 display the data on the process level (X), while Step 7 does so on the empirical, sample level (x).

of parallel computing processes, called threads, and other properties of the calculation. OpenMP is the current standard of such directives; it intervenes at the compiler level, where the source code is transformed into an executable computer program.

E.3 Random Numbers

Random numbers are by definition unknown and unpredictable. They are needed for performing Monte Carlo experiments, where algorithms (e.g., estimation) for stochastic processes (Eq. 1.2) are examined by means of mathematical simulations. In principle, physical experiments with uncontrollable factors can supply the random numbers – for example, temperature measurements on a computer chip. However, the majority of random numbers are generated by mathematical algorithms. Since the properties of these algorithms can be controlled, the produced random numbers can be theoretically predicted. At least, the numbers appear like random. Hence, these tools are often called pseudorandom generators.

A simple form is the multiplicative congruential generator,

$$Z_i = A \times Z_{i-1} \ (\text{mod } M), \ i \geq 1, \tag{E.7}$$

where A, M, and the pseudorandom numbers, Z_i, are integers. That means, Z_i is the remainder obtained when $A \times Z_{i-1}$ is divided by the modulus, M. The first number, Z_1, is primed or seeded. The set of numbers, $\{Z_i\}$, can be mapped onto the interval $[0; 1]$ to produce a uniform distribution, $\mathcal{E}_{\text{U}[0; 1]}(i)$. Ideally, the generator "supplies sequences of numbers from which arbitrarily selected nonoverlapping subsequences appear to behave like statistically independent sequences and where the variation in an arbitrarily chosen subsequence of length k (≥ 1) resembles that of a sample drawn from the uniform distribution on the k-dimensional unit hypercube" (Fishman 1996: p. 587 therein).

A drawback of linear congruential generators, such as that shown in Eq. (E.7), is that generated subsequences may show a certain degree of serial dependence, violating Fishman's ideal. Therefore, in practice, often a second generator is employed, which shuffles the output of the first generator. Nonlinear equations may also be used instead of Eq. (E.7), but these nonlinear generators are slower.

From the uniform distribution over $[0; 1]$, other distributions can be derived by transformations. For example, $\mathcal{E}_{\text{U}[a;b]}(i)$ is from a linear transformation, and $\mathcal{E}_{\text{N}(\mu, \sigma^2)}(i)$ is from the transformation named after Box and Muller (1958).

A simple example – for illustration only – is the linear generator with $A = 7$, $M = 17$, and $Z_1 = 5$, which yields the sequence 5, 1, 7, 15, 3, 4, 11, 9, 12, 16, 10, 2, 14, 13, 6, 8, and 5. An important property of a generator is its period, that is, the maximum length of the sequence before it repeats. In the simple example, the period is only 16.

How can random number generators be effectively used for parallel computing in Monte Carlo experiments (Figure E.2)? Consider first a sequential computing experiment, where the number of threads is equal to one. Here, the only task is to use a good generator to come close to Fishman's ideal. Consider next a number of threads greater than one. Since all threads use the same generator, they have to employ different seeds. Furthermore, the subsequences of integers, Z_i in Eq. (E.7), for all threads must not show coincidences. In the Monte Carlo experiment (Figure E.2), the number of calls of the random number generator, that is, the subsequence length, is equal to the data size, n. This has to be taken into account when selecting the seeds for the threads. The Fortran random number generator MZRAN (Marsaglia and Zaman 1994) has a period of about 10^{28} and avoids many of the mentioned serial dependence problems. An implementation of MZRAN for parallel computing, to which a seed file belongs, is included in the HT software (Section F.2). The number of seeds is 100,000, which may be judged (currently) as allowing for many parallel threads. Each seed leads to subsequence lengths of up to 1,000,000,000 without coincidences, which may be judged as permitting the analysis of large data sizes.

Reading Material

Chandra et al. (2001) is a textbook on parallel programming in OpenMP. Chirila and Lohmann (2015) is a textbook on Fortran programming with a focus on parallel computing and handling of climate model output; it contains examples from physics and climatology. Coles (2001) is a textbook on extreme value analysis with a focus on maximum likelihood estimation. Dahlquist and Björck (2008) is a textbook on numerical techniques. Ellis et al. (1994) is a textbook on Fortran programming. Hager and Wellein (2011) is a textbook on parallel programming with OpenMP and other approaches, which contains exercises and solutions. Michalewicz and Fogel (2000) is a textbook on heuristics and optimization, which contains interesting problems to think about. Press et al. (1996) is a reference book on numerical techniques (also for parallel computing) implemented in Fortran, which contains algorithms; there exist book versions for C/C++.

Box and Muller (1958) is the paper for the Box–Muller transformation for the generation of a Gaussian from a uniform distribution. Marsaglia and Zaman (1994) is a paper on random number generators, which contains code.

Battaglia and Protopapas (2012) is a paper on detecting change-points in temperature series by means of a genetic algorithm. Silver et al. (2016) is an example of the success of heuristics for large search spaces; the described program defeated the world leading human Go player in March 2016.

Appendix F   Data and Software

This appendix lists the datasets analyzed and the software used in this textbook. This allows the replication of the analyses and to do the exercises at the end of Chapters 2 and 3.

The software tools further serve you to carry out own analyses on your data. Data and software can be downloaded from the website for this book. See the preface for the URL.

F.1 Data

The data files required for the exercises are LML.txt (Exercise 2.3), Elbe-S-M1.txt (Exercise 3.2), Elbe-S.txt (Exercise 3.3), and Elbe-W23.txt (Exercise 3.4).

The data files required for the replication of the analyses are named after the figures in which they are shown. For example, the $\delta^{18}O$ data from Figure 1.3 are contained in the file 1_3.txt.

F.2 Software

AMICE. This tool performs the calculation of indices for extremes (Tables 5.1 and 5.2), including the novel action measure and the novel LFSI. In the book, the software is used for the analysis of heatwaves and cold spells (Chapter 5). The source code is included. More information can be found in the source code.

Caliza. This tool performs the occurrence rate estimation and the Cox–Lewis test. In the book, the software is used for the analysis of floods (Chapter 4), heatwaves and cold spells (Chapter 5), and hurricanes and other storms (Chapter 6). The source code is not included. More information can be found in the software manual.

CLsim. This tool performs a sensitivity analysis for the Cox–Lewis test to examine the influence of uncertainties. In the book, the software is used for the analysis of US landfalling hurricanes during the interval from 1980 to 2018 Section 6.1.3). The source code is included. More information can be found in the source code.

GEVMLEST. This tool fits a GEV distribution by means of maximum likelihood estimation. The original routines are from Hosking (1985) and Macleod (1989). In the book, the software is used for the statistical analysis of Elbe summer floods (Chapter 3). The source code is included. More information can be found in the source code.

HT. This tool can be used for the estimation of the heavy-tail index parameter and for the generation of artificial data with that property. In the book, the software is used in an exercise for the analysis of artificial time series and Elbe summer floods (Chapter 3). HT employs parallel computing; it includes an implementation of the random number generator MZRAN. The source code is included. More information can be found in the source code and in the software manual.

TAUEST. This tool performs the estimation of the persistence time on unevenly spaced time series. In the book, the software is used for an autocorrelation analysis of a monsoon series (Chapter 4). The source code is included. More information can be found in the source code.

Appendix G List of Symbols and Abbreviations

Abbreviations

AD	Anno Domini
AOGCM	Atmosphere–Ocean General Circulation Model
ATN|10p|	action measure for cold nights (cold spell index)
ATN|90p|	action measure for warm nights (heatwave index)
ATX|10p|	action measure for cold days (cold spell index)
ATX|90p|	action measure for warm days (heatwave index)
a.u.	arbitrary units
BC	before Christ
BP	before the present
CFC	chlorofluorocarbon
CI	confidence interval
CMIP6	Coupled Model Intercomparison Project Phase 6
CSDI	cold spell duration index
DJF	December (of preceding year)–January–February (season)
ESM	Earth System Model
FD	frost days (cold spell index)
GEOSS	Global Earth Observation System of Systems
GEV (distribution)	Generalized Extreme Value (distribution)
GHG	greenhouse gas
GP (distribution)	Generalized Pareto (distribution)
HQ_{100}	100-year return level for runoff
IPCC	Intergovernmental Panel on Climate Change
ITCZ	Intertropical Convergence Zone
JJA	June–July–August (season)
LFD	Last Frost Day

LFSI	Late Frost Severity Index
LIDAR	Light Detection and Ranging
LWR	longwave radiation
MAD	median of absolute distances to the median
NAO	North Atlantic Oscillation
OLS	ordinary least squares
PDSI	Palmer Drought Severity Index
POT	peaks over threshold
RMSE	root mean squared error
SGS	Start of Growing Season
SI	The International System of Units
	(Le Système international d'unités)
SST	sea-surface temperature
SWR	shortwave radiation
TN10p	cold nights (cold spell index)
TN90p	warm nights (heatwave index)
TNn	coldest minimum daily temperature (cold spell index)
TNx	warmest minimum daily temperature (heatwave index)
TR	tropical nights (heatwave index)
TX10p	cold days (cold spell index)
TX90p	warm days (heatwave index)
TXn	coldest maximum daily temperature (cold spell index)
TXx	warmest maximum daily temperature (heatwave index)
VPDB	Vienna Pee Dee Belemnite
WLS	weighted least squares
WSDI	warm spell duration index

Chemical Substances

CH_4	methane
CO_2	carbon dioxide
N_2O	nitrous oxide
O_3	ozone

SI Prefixes

Y	yotta (10^{24})
Z	zetta (10^{21})
E	exa (10^{18})
P	peta (10^{15})

T	tera (10^{12})
G	giga (10^9)
M	mega (10^6)
k	kilo (10^3)
h	hecto (10^2)
da	deka (10^1)
d	deci (10^{-1})
c	centi (10^{-2})
m	milli (10^{-3})
μ	micro (10^{-6})
n	nano (10^{-9})
p	pico (10^{-12})
f	femto (10^{-15})
a	atto (10^{-18})
z	zepto (10^{-21})
y	yocto (10^{-24})

Base SI Units

A	ampere (electric current)
cd	candela (luminous intensity)
K	kelvin (temperature)
kg	kilogram (mass)
m	meter (length)
mol	mole (amount of substance)
s	second (time)

Derived SI Units

cm	centimeter
dm	decimeter
km	kilometer
μm	micrometer
mg	milligram

Non-SI Units

a	year (annum)
ka	kiloyears (thousand years)
Ma	megayears (million years)

Ga	gigayears (billion years)
d	day
Beaufort	integer unit for wind speed (nonlinear, monotonic relation)
°C	degree Celsius (temperature) (a value expressed in °C equals the value in K plus 273.15)
kn	knot (1 kn = 463/900 m s^{-1})

Mathematical Notation

$\in$	element of
$\approx$	approximately equal to (in equations)
$\neq$	not equal to
$\equiv$	is defined as
$\propto$	proportional to
$\sim$	similar to (in text)
$\mid$	conditional on
$\mid \cdot \mid$	absolute value (of an argument)
$\forall$	for all
$\mathcal{E}_{N(\mu,\,\sigma^2)}(i)$	random variable, normal distribution, mean μ, variance σ^2
$\mathcal{E}_{U[a;b]}(i)$	random variable, uniform distribution, interval $[a;b]$
$\ln(\cdot)$	natural logarithm (of an argument)
mod	modulus
$NINT(\cdot)$	nearest integer function (of an argument)
$\int$	integral
$\sum$	sum
$(\partial f/\partial x)$	partial derivative of a function of several variables, $f(x, y)$, with respect to a certain variable, x
$\nabla(\boldsymbol{x})$	gradient of a vector $\boldsymbol{x}$
$\boldsymbol{x}'$	transpose ("row") of a vector ("column") $\boldsymbol{x}$

References

Allen J. T. (2018) Climate change and severe thunderstorms. In: von Storch H. (Ed.) *Oxford Research Encyclopedia of Climate Science.* Oxford University Press, New York. [doi:10.1093/acrefore/9780190228620.013.62]

Anderegg W. R. L., Konings A. G., Trugman A. T., Yu K., Bowling D. R., Gabbitas R., Karp D. S., Pacala S., Sperry J. S., Sulman B. N., Zenes N. (2018) Hydraulic diversity of forest regulates ecosystem resilience during drought. *Nature* 561(7724): 538–541.

Aristotle (1936) *Aristotle's Physics: A Revised Text with Introduction and Commentary.* Translated and commented by W. D. Ross. Clarendon Press, Oxford, 750 pp.

Battaglia F., Protopapas M. K. (2012) An analysis of global warming in the Alpine region based on nonlinear nonstationary time series models. *Statistical Methods and Applications* 21(3): 315–334.

Bell E. T. (1986) *Men of Mathematics.* First Touchstone edition. Simon & Schuster, New York, 590 pp. [originally published in 1937]

Besonen M. R. (2006) *A 1,000 year high-resolution hurricane history for the Boston area based on the varved sedimentary record from the Lower Mystic Lake (Medford/Arlington, MA).* Ph.D. Dissertation. University of Massachusetts at Amherst, Amherst, MA, 297 pp.

Besonen M. R., Bradley R. S., Mudelsee M., Abbott M. B., Francus P. (2008) A 1,000-year, annually-resolved record of hurricane activity from Boston, Massachusetts. *Geophysical Research Letters* 35(14): L14705. [doi:10.1029/2008GL033950]

Bigler C., Bugmann H. (2018) Climate-induced shifts in leaf unfolding and frost risk of European trees and shrubs. *Scientific Reports* 8(1): 9865. [doi:10.1038/s41598-018-27893-1]

Biswas T. K., Mosley L. M. (2019) From mountain ranges to sweeping plains, in droughts and flooding rains; river Murray water quality over the last four decades. *Water Resources Management* 33(3): 1087–1101.

Blöschl G., Hall J., Viglione A., Perdigão R. A. P., Parajka J., Merz B., Lun D., Arheimer B., Aronica G. T., Bilibashi A., Boháč M., Bonacci O., Borga M., Čanjevac I., Castellarin A., Chirico G. B., Claps P., Frolova N., Ganora D., Gorbachova L., Gül A., Hannaford J., Harrigan S., Kireeva M., Kiss A., Kjeldsen T. R., Kohnová S., Koskela J. J., Ledvinka O., Macdonald N., Mavrova-Guirguinova M., Mediero L., Merz R., Molnar P., Montanari A., Murphy C., Osuch M., Ovcharuk V., Radevski I., Salinas J. L., Sauquet E., Šraj M., Szolgay J., Volpi E., Wilson D., Zaimi K., Živković N. (2019) Changing climate both increases and decreases European river floods. *Nature* 573(7772): 108–111.

Bork H.-R., Bork H., Dalchow C., Faust B., Piorr H.-P., Schatz T. (1998) *Landschaftsentwicklung in Mitteleuropa*. Klett-Perthes, Gotha, 328 pp. [in German]

Box G. E. P. (1953) Non-normality and tests on variances. *Biometrika* 40(3–4): 318–335.

Box G. E. P., Muller M. E. (1958) A note on the generation of random normal deviates. *Annals of Mathematical Statistics* 29(2): 610–611.

Box J. F. (1978) *R.A. Fisher: The Life of a Scientist*. Wiley, New York, 512 pp.

Bradley R. S. (1999) *Paleoclimatology: Reconstructing Climates of the Quaternary*. Second edition. Academic Press, San Diego, 610 pp.

Brázdil R., Glaser R., Pfister C., Dobrovolný P., Antoine J.-M., Barriendos M., Camuffo D., Deutsch M., Enzi S., Guidoboni E., Kotyza O., Rodrigo F. S. (1999) Flood events of selected European rivers in the sixteenth century. *Climatic Change* 43(1): 239–285.

Brockwell P. J., Davis R. A. (1991) *Time Series: Theory and Methods*. Second edition. Springer, New York, 577 pp.

Brockwell P. J., Davis R. A. (1996) *Introduction to Time Series and Forecasting*. Springer, New York, 420 pp.

Brönnimann S., Luterbacher J., Ewen T., Diaz H. F., Stolarski R. S., Neu U. (Eds.) (2008) *Climate Variability and Extremes during the Past 100 Years*. Springer, Dordrecht, Netherlands, 361 pp.

Brooks M. M., Marron J. S. (1991) Asymptotic optimality of the least-squares cross-validation bandwidth for kernel estimates of intensity functions. *Stochastic Processes and their Applications* 38(1): 157–165.

Brückner E. (1890) Klimaschwankungen seit 1700 nebst Bemerkungen über die Klimaschwankungen der Diluvialzeit [Climate variations since 1700 and remarks on climate variations in diluvial time]. *Geographische Abhandlungen* 4(2): 153–484. [in German]

Bryant E. (1991) *Natural Hazards*. Cambridge University Press, Cambridge, 294 pp.

Bundesanstalt für Gewässerkunde (2000) *Untersuchungen zum Abflussregime der Elbe*. Bundesanstalt für Gewässerkunde, Berlin. [BfG-Bericht No. 1228; in German]

Cannell M. G. R., Smith R. I. (1986) Climatic warming, spring budburst and frost damage on trees. *Journal of Applied Ecology* 23(1): 177–191.

Chandra R., Dagum L., Kohr D., Maydan D., McDonald J., Menon R. (2001) *Parallel Programming in OpenMP*. Academic Press, San Diego, 230 pp.

Chatfield C. (2004) *The Analysis of Time Series: An Introduction*. Sixth edition. Chapman & Hall, Boca Raton, FL, 333 pp.

Chirila D. B., Lohmann G. (2015) *Introduction to Modern Fortran for the Earth System Sciences*. Springer, Heidelberg, 250 pp.

Christensen J. H., Krishna Kumar K., Aldrian E., An S.-I., Cavalcanti I. F. A., de Castro M., Dong W., Goswami P., Hall A., Kanyanga J. K., Kitoh A., Kossin J., Lau N.-C., Renwick J., Stephenson D. B., Xie S.-P., Zhou T. (2013) Climate phenomena and their relevance for future regional climate change. In: Stocker T. F., Qin D., Plattner G.-K., Tignor M. M. B., Allen S. K., Boschung J., Nauels A., Xia Y., Bex V., Midgley P. M. (Eds.) *Climate Change 2013: The Physical Science Basis. Working Group I Contribution to the Fifth Assessment Report of the Intergovernmental Panel on Climate Change*. Cambridge University Press, Cambridge, pp. 1217–1308.

Cohen J., Screen J. A., Furtado J. C., Barlow M., Whittleston D., Coumou D., Francis J., Dethloff K., Entekhabi D., Overland J., Jones J. (2014) Recent Arctic amplification and extreme mid-latitude weather. *Nature Geoscience* 7(9): 627–637.

Coles S. (2001) *An Introduction to Statistical Modeling of Extreme Values*. Springer, London, 208 pp.

Conradt T., Roers M., Schröter K., Elmer F., Hoffmann P., Koch H., Hattermann F. F., Wechsung F. (2013) Vergleich der Extremhochwässer 2002 und 2013 im deutschen Teil des Elbegebiets und deren Abflusssimulation durch SWIM-live. *Hydrologie und Wasserbewirtschaftung* 57(5): 241–245. [in German]

Cowling A., Hall P. (1996) On pseudodata methods for removing boundary effects in kernel density estimation. *Journal of the Royal Statistical Society, Series B* 58(3): 551–563.

Cowling A., Hall P., Phillips M. J. (1996) Bootstrap confidence regions for the intensity of a Poisson point process. *Journal of the American Statistical Association* 91(436): 1516–1524.

Cox D. R., Lewis P. A. W. (1966) *The Statistical Analysis of Series of Events.* Methuen, London, 285 pp.

Cronin T. M. (2010) *Paleoclimates: Understanding Climate Change Past and Present.* Columbia University Press, New York, 441 pp.

Dahlquist G., Björck Å. (2008) *Numerical Methods in Scientific Computing*, volume 1, SIAM, Philadelphia, PA, 717 pp. [there exists a second volume, which has not yet been published]

Dansgaard W., Johnsen S. J., Clausen H. B., Dahl-Jensen D., Gundestrup N. S., Hammer C. U., Hvidberg C. S., Steffensen J. P., Sveinbjörnsdottir A. E., Jouzel J., Bond G. (1993) Evidence for general instability of past climate from a 250-kyr ice-core record. *Nature* 364(6434): 218–220.

Delgado S., Landsea C. W., Willoughby H. (2018) Reanalysis of the 1954–63 Atlantic hurricane season. *Journal of Climate* 31(11): 4177–4192.

Deutscher Wetterdienst (2002) *Klimastatusbericht 2001.* Deutscher Wetterdienst, Offenbach am Main, 290 pp. [in German]

Donat M. G., Alexander L. V., Herold N., Dittus A. J. (2016) Temperature and precipitation extremes in century-long gridded observations, reanalyses, and atmospheric model simulations. *Journal of Geophysical Research: Atmospheres* 121(19): 11174–11189.

Donat M. G., Alexander L. V., Yang H., Durre I., Vose R., Caesar J. (2013a) Global land-based datasets for monitoring climatic extremes. *Bulletin of the American Meteorological Society* 94(7): 997–1006.

Donat M. G., Alexander L. V., Yang H., Durre I., Vose R., Dunn R. J. H., Willett K. M., Aguilar E., Brunet M., Caesar J., Hewitson B., Jack C., Klein Tank A. M. G., Kruger A. C., Marengo J., Peterson T. C., Renom M., Oria Rojas C., Rusticucci M., Salinger J., Elrayah A. S., Sekele S. S., Srivastava A. K., Trewin B., Villarroel C., Vincent L. A., Zhai P., Zhang X., Kitching S. (2013b) Updated analyses of temperature and precipitation extreme indices since the beginning of the twentieth century: The HadEX2 dataset. *Journal of Geophysical Research: Atmospheres* 118(5): 2098–2118.

Efron B. (1979) Bootstrap methods: Another look at the jackknife. *The Annals of Statistics* 7(1): 1–26.

Efron B., Hastie T. (2016) *Computer Age Statistical Inference: Algorithms, Evidence, and Data Science.* Cambridge University Press, New York, 475 pp.

Efron B., Tibshirani R. J. (1993) *An Introduction to the Bootstrap*. Chapman & Hall, New York, 436 pp.

Ellis T. M. R., Philips I. R., Lahey T. M. (1994) *Fortran 90 Programming*. Addison-Wesley, Harlow, United Kingdom, 825 pp.

Elsner J. B., Kara A. B. (1999) *Hurricanes of the North Atlantic: Climate and Society*. Oxford University Press, New York, 488 pp.

Emanuel K. A. (1987) The dependence of hurricane intensity on climate. *Nature* 326(6112): 483–485.

Engeln-Müllges G., Reutter F. (1993) *Numerik-Algorithmen mit FORTRAN 77-Programmen*. Seventh edition. BI Wissenschaftsverlag, Mannheim, 1245 pp. [in German]

Eyring V., Bony S., Meehl G. A., Senior C. A., Stevens B., Stouffer R. J., Taylor K. E. (2016) Overview of the Coupled Model Intercomparison Project Phase 6 (CMIP6) experimental design and organization. *Geoscientific Model Delevopment* 9(5): 1937–1958.

Fairchild I. J., Baker A. (2012) *Speleothem Science: From Process to Past Environments*. Wiley-Blackwell, Chichester, 432 pp.

Felis T., McGregor H. V., Linsley B. K., Tudhope A. W., Gagan M. K., Suzuki A., Inoue M., Thomas A. L., Esat T. M., Thompson W. G., Tiwari M., Potts D. C., Mudelsee M., Yokoyama Y., Webster J. M. (2014) Intensification of the meridional temperature gradient in the Great Barrier Reef following the Last Glacial Maximum. *Nature Communications* 5: 4102. [doi:10.1038/ncomms5102]

Fischer E. M., Schär C. (2010) Consistent geographical patterns of changes in high-impact European heatwaves. *Nature Geoscience* 3(6): 398–403.

Fisher R. A. (1915) Frequency distribution of the values of the correlation coefficient in samples from an indefinitely large population. *Biometrika* 10(4): 507–521.

Fisher R. A. (1921) On the "probable error" of a coefficient of correlation deduced from a small sample. *Metron* 1(4): 3–32.

Fisher R. A. (1922) On the mathematical foundations of theoretical statistics. *Philosophical Transactions of the Royal Society of London, Series A* 222: 309–368.

Fisher R. A. (1925) Theory of statistical estimation. *Mathematical Proceedings of the Cambridge Philosophical Society* 22(5): 700–725.

Fisher R. A. (1929) Tests of significance in harmonic analysis. *Proceedings of the Royal Society of London, Series A* 125(796): 54–59.

Fisher R. A., Tippett L. H. C. (1928) Limiting forms of the frequency distribution of the largest or smallest member of a sample. *Mathematical Proceedings of the Cambridge Philosophical Society* 24(2): 180–190.

Fishman G. S. (1996) *Monte Carlo: Concepts, Algorithms, and Applications.* Springer, New York, 698 pp.

Flato G., Marotzke J., Abiodun B., Braconnot P., Chou S. C., Collins W., Cox P., Driouech F., Emori S., Eyring V., Forest C., Gleckler P., Guilyardi E., Jakob C., Kattsov V., Reason C., Rummukainen M. (2013) Evaluation of climate models. In: Stocker T. F., Qin D., Plattner G.-K., Tignor M. M. B., Allen S. K., Boschung J., Nauels A., Xia Y., Bex V., Midgley P. M. (Eds.) *Climate Change 2013: The Physical Science Basis. Working Group I Contribution to the Fifth Assessment Report of the Intergovernmental Panel on Climate Change.* Cambridge University Press, Cambridge, pp. 741–866.

Fleitmann D., Burns S. J., Mangini A., Mudelsee M., Kramers J., Villa I., Neff U., Al-Subbary A. A., Buettner A., Hippler D., Matter A. (2007) Holocene ITCZ and Indian monsoon dynamics recorded in stalagmites from Oman and Yemen (Socotra). *Quaternary Science Reviews* 26(1–2): 170–188.

Fleitmann D., Burns S. J., Mudelsee M., Neff U., Kramers J., Mangini A., Matter A. (2003) Holocene forcing of the Indian monsoon recorded in a stalagmite from southern Oman. *Science* 300(5626): 1737–1739.

Fohlmeister J., Scholz D., Kromer B., Mangini A. (2011) Modelling carbon isotopes of carbonates in cave drip water. *Geochimica et Cosmochimica Acta* 75(18): 5219–5228.

Folland C. K., Miller C., Bader D., Crowe M., Jones P., Plummer N., Richman M., Parker D. E., Rogers J., Scholefield P. (1999) Workshop on Indices and Indicators for Climate Extremes, Asheville, NC, USA, 3–6 June 1997 – Breakout Group C: Temperature indices for climate extremes. *Climatic Change* 42(1): 31–43.

Forster P., Ramaswamy V., Artaxo P., Berntsen T., Betts R., Fahey D. W., Haywood J., Lean J., Lowe D. C., Myhre G., Nganga J., Prinn R., Raga G., Schulz M., Van Dorland R. (2007) Changes in atmospheric constituents and in radiative forcing. In: Solomon S., Qin D., Manning M., Marquis M., Averyt K., Tignor M. M. B., Miller Jr H. L., Chen Z. (Eds.) *Climate Change 2007: The Physical Science Basis. Contribution of Working Group I to the Fourth Assessment Report of the Intergovernmental Panel on Climate Change.* Cambridge University Press, Cambridge, pp. 129–234.

Fouillet A., Rey G., Laurent F., Pavillon G., Bellec S., Guihenneuc-Jouyaux C., Clavel J., Jougla E., Hémon D. (2006) Excess mortality related to the August 2003 heat wave in France. *International Archives of Occupational and Environmental Health* 80(1): 16–24.

Giffard-Roisin S., Yang M., Charpiat G., Kégl B., Monteleoni C. (2018) Fused deep learning for hurricane track forecast from reanalysis data. 8th International Workshop on Climate Informatics, National Center for Atmospheric Research, Boulder, CO, 19 to 21 September 2018, 4 pp.

Giorgi F., Gao X.-J. (2018) Regional earth system modeling: Review and future directions. *Atmospheric and Oceanic Science Letters* 11(2): 189–197.

Gradstein F. M., Ogg J. G., Smith A. G. (Eds.) (2004) *A Geologic Time Scale 2004*. Cambridge University Press, Cambridge, 589 pp.

Hager G., Wellein G. (2011) *Introduction to High Performance Computing for Scientists and Engineers*. CRC Press, Boca Raton, FL, 330 pp.

Hammond J. M. (1990) Storm in a teacup or winds of change? *Weather* 45(12): 443–448.

Hartmann D. L., Klein Tank A. M. G., Rusticucci M., Alexander L. V., Brönnimann S., Charabi Y. A.-R., Dentener F. J., Dlugokencky E. J., Easterling D. R., Kaplan A., Soden B. J., Thorne P. W., Wild M., Zhai P. (2013) Observations: Atmosphere and surface. In: Stocker T. F., Qin D., Plattner G.-K., Tignor M. M. B., Allen S. K., Boschung J., Nauels A., Xia Y., Bex V., Midgley P. M. (Eds.) *Climate Change 2013: The Physical Science Basis. Working Group I Contribution to the Fifth Assessment Report of the Intergovernmental Panel on Climate Change*. Cambridge University Press, Cambridge, pp. 159–254.

Hill B. M. (1975) A simple general approach to inference about the tail of a distribution. *The Annals of Statistics* 3(5): 1163–1174.

Hosking J. R. M. (1985) Maximum-likelihood estimation of the parameters of the Generalized Extreme-Value distribution. *Applied Statistics* 34(3): 301–310.

Hübener W. (1983) Occam's Razor not mysterious. *Archiv für Begriffsgeschichte* 27: 73–92. [in German, with excerpts in English, French, Greek, and Latin]

Hurrell J. W. (1995) Decadal trends in the North Atlantic Oscillation: Regional temperatures and precipitation. *Science* 269(5524): 676–679.

Huxham J. (1758) An account of the extraordinary heat of the weather in July 1757, and of the effects of it. *Philosophical Transactions of the Royal Society of London* 50(2): 523–524.

Ivanovich M., Harmon R. S. (Eds.) (1992) *Uranium-Series Disequilibrium: Applications to Earth, Marine, and Environmental Sciences*. Second edition. Clarendon Press, Oxford, 910 pp.

Jílek P., Melková J., Růžičková E., Šilar J., Zeman A. (1995) Radiocarbon dating of Holocene sediments: Flood events and evolution of the Labe (Elbe) river in central Bohemia (Czech Republic). *Radiocarbon* 37(2): 131–137.

Johnson N. L., Kotz S., Balakrishnan N. (1994) *Continuous Univariate Distributions*, volume 1. Second edition. Wiley, New York, 756 pp.

Johnson N. L., Kotz S., Balakrishnan N. (1995) *Continuous Univariate Distributions*, volume 2. Second edition. Wiley, New York, 719 pp.

Jun M., Knutti R., Nychka D. W. (2008) Spatial analysis to quantify numerical model bias and dependence: How many climate models are there? *Journal of the American Statistical Association* 103(483): 934–947.

Kendall M. G. (1954) Note on bias in the estimation of autocorrelation. *Biometrika* 41(3–4): 403–404.

Kirtman B., Power S. B., Adedoyin A. J., Boer G. J., Bojariu R., Camilloni I., Doblas-Reyes F., Fiore A. M., Kimoto M., Meehl G., Prather M., Sarr A., Schär C., Sutton R., van Oldenborgh G. J., Vecchi G., Wang H.-J. (2013) Near-term climate change: Projections and predictability. In: Stocker T. F., Qin D., Plattner G.-K., Tignor M. M. B., Allen S. K., Boschung J., Nauels A., Xia Y., Bex V., Midgley P. M. (Eds.) *Climate Change 2013: The Physical Science Basis. Working Group I Contribution to the Fifth Assessment Report of the Intergovernmental Panel on Climate Change*. Cambridge University Press, Cambridge, pp. 953–1028.

Klein Tank A. M. G., Wijngaard J. B., Können G. P., Böhm R., Demarée G., Gocheva A., Mileta M., Pashiardis S., Hejkrlik L., Kern-Hansen C., Heino R., Bessemoulin P., Müller-Westermeier G., Tzanakou M., Szalai S., Pálsdóttir T., Fitzgerald D., Rubin S., Capaldo M., Maugeri M., Leitass A., Bukantis A., Aberfeld R., van Engelen A. F. V., Forland E., Mietus M., Coelho F., Mares C., Razuvaev V., Nieplova E., Cegnar T., López J. A., Dahlström B., Moberg A., Kirchhofer W., Ceylan A., Pachaliuk O., Alexander L. V., Petrovic P. (2002) Daily dataset of 20th-century surface air temperature and precipitation series for the European Climate Assessment. *International Journal of Climatology* 22(12): 1441–1453.

Knabb R. D., Rhome J. R., Brown D. P. (2011) *Tropical Cyclone Report: Hurricane Katrina, 23–30 August 2005*. National Oceanic and Atmospheric Administration, National Hurricane Center, Miami, 43 pp.

Königliche Elbstrom-Bauverwaltung (1893) *Hydrologischer Jahresbericht von der Elbe für 1892 [Hydrological annual report of the Elbe for 1892].* Baensch, Magdeburg, 296 pp. [in German]

Königliche Elbstrombauverwaltung (1898) *Der Elbstrom, sein Stromgebiet und seine wichtigsten Nebenflüsse*, volume 3.1. Dietrich Reimer, Berlin, 436 pp. [in German]

Körber H.-G. (1993) *Die Geschichte des Meteorologischen Observatoriums Potsdam [The history of the meteorological observatory Potsdam].* Deutscher Wetterdienst, Offenbach am Main, 129 pp. [in German]

Kropp J. P., Schellnhuber H. J. (Eds.) (2011) *In Extremis: Disruptive Events and Trends in Climate and Hydrology.* Springer, Berlin, 320 pp.

Kuhn T. S. (1970) *The Structure of Scientific Revolutions.* Second edition. University of Chicago Press, Chicago, 210 pp.

Landsea C. W., Anderson C., Charles N., Clark G., Dunion J., Fernández-Partagás J., Hungerford P., Neumann C., Zimmer M. (2004) The Atlantic Hurricane Database Reanalysis Project: Documentation for 1851–1910 alterations and additions to the HURDAT database. In: Murnane R. J., Liu K.-b. (Eds.) *Hurricanes and Typhoons: Past, Present, and Future.* Columbia University Press, New York, pp. 177–221.

Luterbacher J., Rickli R., Xoplaki E., Tinguely C., Beck C., Pfister C., Wanner H. (2001) The late Maunder Minimum (1675–1715) – A key period for studying decadal scale climatic change in Europe. *Climatic Change* 49(4): 441–462.

Ma Q., Huang J.-G., Hänninen H., Berninger F. (2018) Divergent trends in the risk of spring frost damage to trees in Europe with recent warming. *Global Change Biology* 25(1): 351–360.

Macleod A. J. (1989) A remark on algorithm AS 215: Maximum-likelihood estimation of the parameters of the Generalized Extreme-Value distribution. *Applied Statistics* 38(1): 198–199.

Manley G. (1974) Central England temperatures: Monthly means 1659 to 1973. *Quarterly Journal of the Royal Meteorological Society* 100(425): 389–405.

Mann M. E., Woodruff J. D., Donnelly J. P., Zhang Z. (2009) Atlantic hurricanes and climate over the past 1,500 years. *Nature* 460(7257): 880–883.

Marron J. S. (1988) Automatic smoothing parameter selection: A survey. *Empirical Economics* 13(3–4): 187–208.

Marsaglia G., Zaman A. (1994) Some portable very-long-period random number generators. *Computers in Physics* 8(1): 117–121.

Masson-Delmotte V., Schulz M., Abe-Ouchi A., Beer J., Ganopolski A., González Rouco J. F., Jansen E., Lambeck K., Luterbacher J., Naish T., Osborn T., Otto-Bliesner B., Quinn T., Ramesh R., Rojas M., Shao X., Timmermann A. (2013) Information from paleoclimate archives. In: Stocker T. F., Qin D., Plattner G.-K., Tignor M. M. B., Allen S. K., Boschung J., Nauels A., Xia Y., Bex V., Midgley P. M. (Eds.) *Climate Change 2013: The Physical Science Basis. Working Group I Contribution to the Fifth Assessment Report of the Intergovernmental Panel on Climate Change*. Cambridge University Press, Cambridge, pp. 383–464.

Meier M., Fuhrer J., Holzkämper A. (2018) Changing risk of spring frost damage in grapevines due to climate change? A case study in the Swiss Rhone valley. *International Journal of Biometeorology* 62(6): 991–1002.

Michalewicz Z., Fogel D. B. (2000) *How to Solve It: Modern Heuristics*. Springer, Berlin, 467 pp.

Militzer S. (1998) *Klima, Umwelt, Mensch (1500–1800): Studien und Quellen zur Bedeutung von Klima und Witterung in der vorindustriellen Gesellschaft [Climate, environment, man (1500–1800): Studies and sources on the relevance of climate and weather in the preindustrial society]*, volume 1–3. University of Leipzig, Leipzig, 1971 pp. [in German]

Montgomery D. C., Peck E. A. (1992) *Introduction to Linear Regression Analysis*. Second edition. Wiley, New York, 527 pp.

Mudelsee M. (2000) Ramp function regression: A tool for quantifying climate transitions. *Computers and Geosciences* 26(3): 293–307.

Mudelsee M. (2002) TAUEST: A computer program for estimating persistence in unevenly spaced weather/climate time series. *Computers and Geosciences* 28(1): 69–72.

Mudelsee M. (2014) *Climate Time Series Analysis: Classical Statistical and Bootstrap Methods*. Second edition. Springer, Cham, Switzerland, 454 pp.

Mudelsee M. (2019) Trend analysis of climate time series: A review of methods. *Earth-Science Reviews* 190: 310–322. [doi:10.1016/j.earscirev.2018.12.005]

Mudelsee M., Bermejo M. A. (2017) Optimal heavy tail estimation – Part 1: Order selection. *Nonlinear Processes in Geophysics* 24(4): 737–744.

Mudelsee M., Bickert T., Lear C. H., Lohmann G. (2014) Cenozoic climate changes: A review based on time series analysis of marine benthic $\delta^{18}O$ records. *Reviews of Geophysics* 52(3): 333–374.

Mudelsee M., Börngen M., Tetzlaff G., Grünewald U. (2003) No upward trends in the occurrence of extreme floods in central Europe. *Nature* 425(6954): 166–169.

Mudelsee M., Börngen M., Tetzlaff G., Grünewald U. (2004) Extreme floods in central Europe over the past 500 years: Role of cyclone pathway "Zugstrasse Vb." *Journal of Geophysical Research* 109(D23): D23101. [doi:10.1029/2004JD005034]

Mudelsee M., Dcutsch M., Börngen M., Tetzlaff G. (2006) Trends in flood risk of the river Werra (Germany) over the past 500 years. *Hydrological Sciences Journal* 51(5): 818–833.

Munoz S. E., Giosan L., Therrell M. D., Remo J. W. F., Shen Z., Sullivan R. M., Wiman C., O'Donnell M., Donnelly J. P. (2018) Climatic control of Mississippi river flood hazard amplified by river engineering. *Nature* 556(7699): 95–98.

National Academies of Sciences, Engineering, and Medicine (2016) *Attribution of Extreme Weather Events in the Context of Climate Change.* The National Academies Press, Washington, DC, 165 pp.

Neuendorf K. K. E., Mehl Jr J. P., Jackson J. A. (Eds.) (2005) *Glossary of Geology.* Fifth edition. American Geological Institute, Alexandria, VA, 779 pp.

Newton I. (1687) *Philosophiae Naturalis Principia Mathematica.* Jussu Societatis Regiae ac Typis Josephi Streater, London, 510 pp. [In Latin; the sentence cited in our Chapter 5 is from p. 384 of the translation by Andrew Motte from the first American edition, published in 1846 by Daniel Adee, New York.]

Nolan J. P. (2003) Modeling financial data with stable distributions. In: Rachev S. T. (Ed.) *Handbook of Heavy Tailed Distributions in Finance.* Elsevier, Amsterdam, pp. 106–130.

Ocean Drilling Program (1988–2007) *Proceedings of the Ocean Drilling Program, Scientific Results*, volume 101–210. Ocean Drilling Program, College Station, TX.

Orth R., Vogel M. M., Luterbacher J., Pfister C., Seneviratne S. I. (2016) Did European temperatures in 1540 exceed present-day records? *Environmental Research Letters* 11(11): 114021. [doi:10.1088/1748-9326/11/11/114021]

Otto F. E. L., Massey N., van Oldenborgh G. J., Jones R. G., Allen M. R. (2012) Reconciling two approaches to attribution of the 2010 Russian heat wave. *Geophysical Research Letters* 39(4): L04702. [doi:10.1029/2011GL050422]

Peixoto J. P., Oort A. H. (1992) *Physics of Climate*. American Institute of Physics, New York, 520 pp.

Perkins S. E., Alexander L. V. (2013) On the measurement of heat waves. *Journal of Climate* 26(13): 4500–4517.

Pielke Jr R. A., Gratz J., Landsea C. W., Collins D., Saunders M. A., Musulin R. (2008) Normalized hurricane damage in the United States: 1900–2005. *Natural Hazards Review* 9(1): 29–42.

Pierce C. H. (1939) The meteorological history of the New England Hurricane of Sept. 21, 1938. *Monthly Weather Review* 67(8): 237–285.

Pierrehumbert R. T. (2010) *Principles of Planetary Climate*. Cambridge University Press, Cambridge, 652 pp.

Pinto J. G., Reyers M. (2017) Winde und Zyklonen. In: Brasseur G., Jacob D., Schuck-Zöller S. (Eds.) *Klimawandel in Deutschland: Entwicklung, Folgen, Risiken und Perspektiven*. Springer, Berlin, pp. 67–75. [in German]

Pirazzoli P. A., Tomasin A. (2003) Recent near-surface wind changes in the central Mediterranean and Adriatic areas. *International Journal of Climatology* 23(8): 963–973.

Politis D. N., Romano J. P. (1994) The stationary bootstrap. *Journal of the American Statistical Association* 89(428): 1303–1313.

Popper K. R. (1959) *The logic of scientific discovery*. Basic Books, New York, 480 pp. [The original German version was published in 1934.]

Popper K. R. (2004) Woran glaubt der Westen? (gestohlen vom Autor der "Offenen Gesellschaft") [What does the West believe in? (stolen from the author of "The Open Society")]. In: *Auf der Suche nach einer besseren Welt: Vorträge und Aufsätze aus dreißig Jahren*, 13th edition. Piper, Munich, pp. 231–253. [In German; the English translation of the book was published as "In Search of a Better World: Lectures and Essays from Thirty Years" in 1992 by Routledge.]

Press W. H., Teukolsky S. A., Vetterling W. T., Flannery B. P. (1996) *Numerical Recipes in Fortran 90: The Art of Parallel Scientific Computing*. Second edition. Cambridge University Press, Cambridge, pp. 935–1486.

Rahmstorf S., Coumou D. (2011) Increase of extreme events in a warming world. *Proceedings of the National Academy of Sciences of the United States of America* 108(44): 17905–17909. [Correction: 109(12): 4708]

Rappaport E. N., Fernández-Partagás J. J. (1997) History of the deadliest Atlantic tropical cyclones since the discovery of the New World. In: Diaz H. F., Pulwarty R. S. (Eds.) *Hurricanes: Climate and Socioeconomic Impacts*. Springer, Berlin, pp. 93–108.

Reimer P. J., Bard E., Bayliss A., Beck J. W., Blackwell P. G., Ramsey C. B., Buck C. E., Cheng H., Edwards R. L., Friedrich M., Grootes P. M., Guilderson T. P., Haflidason H., Hajdas I., Hatté C., Heaton T. J., Hoffmann D. L., Hogg A. G., Hughen K. A., Kaiser K. F., Kromer B., Manning S. W., Niu M., Reimer R. W., Richards D. A., Scott E. M., Southon J. R., Staff R. A., Turney C. S. M., van der Plicht J. (2013) IntCal13 and Marine13 radiocarbon age calibration curves 0–50,000 years cal BP. *Radiocarbon* 55(4): 1869–1887.

Resnick S. I. (2007) *Heavy-Tail Phenomena: Probabilistic and Statistical Modeling*. Springer, New York, 404 pp.

Röthlisberger R., Bigler M., Hutterli M., Sommer S., Stauffer B., Junghans H. G., Wagenbach D. (2000) Technique for continuous high-resolution analysis of trace substances in firn and ice cores. *Environmental Science & Technology* 34(2): 338–342.

Rubin C. M., Horton B. P., Sieh K., Pilarczyk J. E., Daly P., Ismail N., Parnell A. C. (2017) Highly variable recurrence of tsunamis in the 7,400 years before the 2004 Indian Ocean tsunami. *Nature Communications* 8: 16019. [doi:10.1038/ncomms16019]

Russell B. (1996) *History of Western Philosophy*. Second edition. Routledge, London, 842 pp. [originally published in 1961 by George Allen and Unwin]

Schiesser H. H., Pfister C., Bader J. (1997) Winter storms in Switzerland north of the Alps 1864/1865–1993/1994. *Theoretical and Applied Climatology* 58(1–2): 1–19.

Schmidt M. (2000) *Hochwasser und Hochwasserschutz in Deutschland vor 1850 [Floods and Flood Protection in Germany before 1850]*. Oldenbourg Industrieverlag, Munich, 330 pp. [in German]

Scott D. W. (1979) On optimal and data-based histograms. *Biometrika* 66(3): 605–610.

Seager R., Hoerling M., Schubert S., Wang H., Lyon B., Kumar A., Nakamura J., Henderson N. (2015) Causes of the 2011–14 California drought. *Journal of Climate* 28(18): 6997–7024.

Seibold E., Berger W. H. (1996) *The Sea Floor: An Introduction to Marine Geology*. Third edition. Springer, Berlin, 356 pp.

Sheffield J., Wood E. F., Roderick M. L. (2012) Little change in global drought over the past 60 years. *Nature* 491(7424): 435–438.

Silva A. T. (2017) Introduction to nonstationary analysis and modeling of hydrologic variables. In: Naghettini M. (Ed.) *Fundamentals of Statistical Hydrology*. Springer, Cham, Switzerland, pp. 537–577.

Silver D., Huang A., Maddison C. J., Guez A., Sifre L., van den Driessche G., Schrittwieser J., Antonoglou I., Panneershelvam V., Lanctot M., Dieleman S., Grewe D., Nham J., Kalchbrenner N., Sutskever I., Lillicrap T., Leach M., Kavukcuoglu K., Graepel T., Hassabis D. (2016) Mastering the game of Go with deep neural networks and tree search. *Nature* 529(7587): 484–489.

Sinha A., Stott L., Berkelhammer M., Cheng H., Edwards R. L., Buckley B., Aldenderfer M., Mudelsee M. (2011) A global context for megadroughts in monsoon Asia during the past millennium. *Quaternary Science Reviews* 30(1–2): 47–62.

Smits A., Klein Tank A. M. G., Können G. P. (2005) Trends in storminess over the Netherlands, 1962–2002. *International Journal of Climatology* 25(10): 1331–1344.

Starkel L. (2001) Extreme rainfalls and river floods in Europe during the last millennium. *Geographia Polonica* 74(2): 69–79.

Stocker T. F., Qin D., Plattner G.-K., Tignor M. M. B., Allen S. K., Boschung J., Nauels A., Xia Y., Bex V., Midgley P. M. (Eds.) (2013) *Climate Change 2013: The Physical Science Basis. Working Group I Contribution to the Fifth Assessment Report of the Intergovernmental Panel on Climate Change.* Cambridge University Press, Cambridge, 1535 pp.

Swain D. L., Langenbrunner B., Neelin J. D., Hall A. (2018) Increasing precipitation volatility in twenty-first-century California. *Nature Climate Change* 8(5): 427–433.

Taylor R. E. (1987) *Radiocarbon Dating: An Archaeological Perspective.* Academic Press, Orlando, FL, 212 pp.

Ting M., Kossin J. P., Camargo S. J., Li C. (2019) Past and future hurricane intensity change along the U.S. East Coast. *Scientific Reports* 9(1): 7795. [doi:10.1038/s41598-019-44252-w]

Tukey J. W. (1977) *Exploratory Data Analysis.* Addison-Wesley, Reading, MA, 688 pp.

van der Ploeg R. R., Schweigert P. (2001) Elbe river flood peaks and postwar agricultural land use in East Germany. *Naturwissenschaften* 88(12): 522–525.

Vaughan D. G., Comiso J. C., Allison I., Carrasco J., Kaser G., Kwok R., Mote P., Murray T., Paul F., Ren J., Rignot E., Solomina O., Steffen K., Zhang T. (2013) Observations: Cryosphere. In: Stocker T. F., Qin D., Plattner G.-K., Tignor M. M. B., Allen S. K., Boschung J., Nauels A., Xia Y., Bex V., Midgley P. M. (Eds.) *Climate Change 2013: The Physical Science Basis. Working Group I Contribution to the Fifth Assessment*

Report of the Intergovernmental Panel on Climate Change. Cambridge University Press, Cambridge, pp. 317–382.

Vautard R., van Oldenborgh G. J., Otto F. E. L., Yiou P., de Vries H., van Meijgaard E., Stepek A., Soubeyroux J.-M., Philip S., Kew S. F., Costella C., Singh R., Tebaldi C. (2019) Human influence on European winter storms such as those of January 2018. *Earth System Dynamics* 10(2): 271–286.

von Storch H., Zwiers F. W. (1999) *Statistical Analysis in Climate Research.* Cambridge University Press, Cambridge, 484 pp.

Walsh K. J. E., McBride J. L., Klotzbach P. J., Balachandran S., Camargo S. J., Holland G., Knutson T. R., Kossin J. P., Lee T.-c., Sobel A., Sugi M. (2016) Tropical cyclones and climate change. *Wiley Interdisciplinary Reviews: Climate Change* 7(1): 65–89.

Wang Y., Cheng H., Edwards R. L., He Y., Kong X., An Z., Wu J., Kelly M. J., Dykoski C. A., Li X. (2005) The Holocene Asian monsoon: Links to solar changes and North Atlantic climate. *Science* 308(5723): 854–857.

Warmund M. R., Guinan P., Fernandez G. (2008) Temperatures and cold damage to small fruit crops across the eastern United States associated with the April 2007 freeze. *HortScience* 43(6): 1643–1647.

Webster P. J., Magaña V. O., Palmer T. N., Shukla J., Tomas R. A., Yanai M., Yasunari T. (1998) Monsoons: Processes, predictability, and the prospects for prediction. *Journal of Geophysical Research* 103(C7): 14451–14510.

Weikinn C. (1958) *Quellentexte zur Witterungsgeschichte Europas von der Zeitwende bis zum Jahre 1850: Hydrographie, Teil 1 (Zeitwende–1500) [Source texts on the weather history in Europe from AD 0 to the year 1850: Hydrography, Part 1 (AD 0–1500)].* Akademie-Verlag, Berlin, 531 pp. [in German]

Weikinn C. (1960) *Quellentexte zur Witterungsgeschichte Europas von der Zeitwende bis zum Jahre 1850: Hydrographie, Teil 2 (1501–1600).* Akademie-Verlag, Berlin, 486 pp. [in German]

Weikinn C. (1961) *Quellentexte zur Witterungsgeschichte Europas von der Zeitwende bis zum Jahre 1850: Hydrographie, Teil 3 (1601–1700).* Akademie-Verlag, Berlin, 586 pp. [in German]

Weikinn C. (1963) *Quellentexte zur Witterungsgeschichte Europas von der Zeitwende bis zum Jahre 1850: Hydrographie, Teil 4 (1701–1750).* Akademie-Verlag, Berlin, 381 pp. [in German]

Weikinn C. (2000) *Quellentexte zur Witterungsgeschichte Europas von der Zeitwende bis zum Jahr 1850: Hydrographie, Teil 5 (1751–1800).* Gebrüder Borntraeger, Berlin, 674 pp. [in German]

Weikinn C. (2002) *Quellentexte zur Witterungsgeschichte Europas von der Zeitwende bis zum Jahr 1850: Hydrographie, Teil 6 (1801–1850).* Gebrüder Borntraeger, Berlin, 728 pp. [in German]

Weinkle J., Landsea C., Collins D., Musulin R., Crompton R. P., Klotzbach P. J., Pielke Jr R. (2018) Normalized hurricane damage in the continental United States 1900–2017. *Nature Sustainability* 1(12): 808–813.

Werner M., Langebroek P. M., Carlsen T., Herold M., Lohmann G. (2011) Stable water isotopes in the ECHAM5 general circulation model: Toward high-resolution isotope modeling on a global scale. *Journal of Geophysical Research* 116(D15): D15109. [doi:10.1029/2011JD015681]

Wetter O., Pfister C., Werner J. P., Zorita E., Wagner S., Seneviratne S. I., Herget J., Grünewald U., Luterbacher J., Alcoforado M.-J., Barriendos M., Bieber U., Brázdil R., Burmeister K. H., Camenisch C., Contino A., Dobrovolný P., Glaser R., Himmelsbach I., Kiss A., Kotyza O., Labbé T., Limanówka D., Litzenburger L., Nordli Ø., Pribyl K., Retsö D., Riemann D., Rohr C., Siegfried W., Söderberg J., Spring J.-L. (2014) The year-long unprecedented European heat and drought of 1540 – a worst case. *Climatic Change* 125(3–4): 349–363.

Wheatley J. J., Blackwell P. G., Abram N. J., McConnell J. R., Thomas E. R., Wolff E. W. (2012) Automated ice-core layer-counting with strong univariate signals. *Climate of the Past* 8(6): 1869–1879.

Wilhelm B., Ballestero Cánovas J. A., Macdonald N., Toonen W. H. J., Baker V., Barriendos M., Benito G., Brauer A., Corella J. P., Denniston R., Glaser R., Ionita M., Kahle M., Liu T., Luetscher M., Macklin M., Mudelsee M., Munoz S., Schulte L., St. George S., Stoffel M., Wetter O. (2019) Interpreting historical, botanical, and geological evidence to aid preparations for future floods. *Wiley Interdisciplinary Reviews: Water* 6(1): e1318. [doi:10.1002/wat2.1318]

Winter H. C., Tawn J. A. (2016) Modelling heatwaves in central France: A case-study in extremal dependence. *Applied Statistics* 65(3): 345–365.

Witze A. (2018) The cruellest seas: Extreme floods will become more common as sea levels rise. *Nature* 555(7695): 156–158. [not peer-reviewed]

Zhang X., Alexander L., Hegerl G. C., Jones P., Klein Tank A., Peterson T. C., Trewin B., Zwiers F. W. (2011) Indices for monitoring changes in extremes based on daily temperature and precipitation data. *Wiley Interdisciplinary Reviews: Climate Change* 2(6): 851–870.

Zhang X., Hegerl G., Zwiers F. W., Kenyon J. (2005) Avoiding inhomogeneity in percentile-based indices of temperature extremes. *Journal of Climate* 18(11): 1641–1651.

Index

Italic page numbers refer to figures or tables.

195